I0882123

Ibn ‘Arabî – Time and Cosmology

Time is one of the most important issues in physics, cosmology, philosophy and theology. Many books and articles have been published in this interdisciplinary field before, but none of those studies have fully described Ibn ‘Arabî’s unique view which is central to understanding, for example, his controversial theory of the ‘oneness of being’.

This book is the first comprehensive attempt to explain Ibn ‘Arabî’s distinctive view of time and its role in the process of creating the cosmos and its relation with the Creator. By comparing this original view with modern theories of physics and cosmology, Mohamed Haj Yousef constructs a new cosmological model that may deepen and extend our understanding of the world, while potentially solving some of the drawbacks in the current models such as the historical Zeno’s paradoxes of motion and the recent Einstein–Podolsky–Rosen paradox (EPR) that underlines the discrepancies between Quantum Mechanics and Relativity.

Ibn ‘Arabî – Time and Cosmology is an important contribution to the fields of philosophy, cosmology, physics and theology.

Mohamed Haj Yousef is lecturer in the Department of Physics, United Arab Emirates University.

Culture and civilization in the Middle East

General Editor: Ian Richard Netton

Professor of Islamic Studies, University of Exeter

This series studies the Middle East through the twin foci of its diverse cultures and civilizations. Comprising original monographs as well as scholarly surveys, it covers topics in the fields of Middle Eastern literature, archaeology, law, history, philosophy, science, folklore, art, architecture and language. While there is a plurality of views, the series presents serious scholarship in a lucid and stimulating fashion.

1 **The Epistemology of Ibn Khaldun**
Zaid Ahmad

2 **The Hanbali School of Law and Ibn Taymiyyah**
Conflict or conciliation
Abdul Hakim I Al-Matroudi

3 **Arabic Rhetoric**
A pragmatic analysis
Hussein Abdul-Raof

4 **Arab Representations of the Occident**
East-West encounters in Arabic fiction
Rasheed El-Enany

5 **God and Humans in Islamic Thought**
Abd al-Jabbār, Ibn Sīnā and al-Ghazālī
Maha Elkaisy-Friemuth

6 **Original Islam**
Malik and the madhhab of Madina
Yasin Dutton

7 **Al-Ghazali and the Qur'an**
One book, many meanings
Martin Whittingham

Ibn ‘Arabî – Time and Cosmology

Mohamed Haj Yousef

Routledge
Taylor & Francis Group
LONDON AND NEW YORK

First published 2008
by Routledge
2 Park Square, Milton Park, Abingdon, Oxon, OX14 4RN

Simultaneously published in the USA and Canada
by Routledge
270 Madison Ave, New York NY 10016

Routledge is an imprint of the Taylor & Francis Group, an informa business

First issued in paperback 2011

Typeset in Times by Wearset Ltd, Boldon, Tyne and Wear

British Library Cataloguing in Publication Data
A catalogue record for this book is available from the British Library

Library of Congress Cataloging in Publication Data
A catalog record for this book has been requested

ISBN 10: 0-415-44499-3 (hbk)
ISBN 10: 0-415-66401-2 (pbk)
ISBN 10: 0-203-93824-0 (ebk)

ISBN 13: 978-0-415-44499-6 (hbk)
ISBN 13: 978-0-415-66401-1 (pbk)
ISBN 13: 978-0-203-93824-9 (ebk)

To the Spirit of al-Shaykh al-Akbar Muhyi ed-Din Ibn al-'Arabî, may Allah's Mercy be upon him,
To the Spirits of my parents, may Allah's Mercy be upon them,
To my dearest wife who supported me all the times,
To my two sons Abdullah and Yousef, and my daughter Fatima,
To my brothers and sisters, in Islam, and in humankind,
To all I dedicate this humble work.

Contents

Figures

Tables

Foreword

Students of the world's religious traditions, together with specialists in the history of premodern science and philosophy, are well aware of the centrality within the scriptures and theologies of the major world religions, over many centuries, of detailed symbolic accounts of cosmology and metaphysics (including the intricate problematics of creation) – and of the crucial role played within each of those religious traditions by corresponding philosophical and scientific schemas of astronomy and cosmology that often provided a common language and framework of understanding shared by their educated elites. In premodern times, this key interpretative function was particularly important in the case of that complex of Hellenistic philosophic and cosmological disciplines largely shared by educated proponents of each of the three Abrahamic faiths. Given today's widespread journalistic stereotypes about the supposed 'opposition' of science and religion, this book is a salutary reminder – and an extraordinarily rich and detailed illustration – of the complex interpenetration of philosophical and scriptural elements throughout the central traditions of later Islamic thought, prior to the recent scientific revolutions. At the same time, Dr Haj Yousef's training and expertise as a modern physicist allow him to suggest, in his provocative final chapter, intriguing ways in which the earlier cosmological and theological speculations of Ibn 'Arabî carefully outlined in this study may also parallel very recent developments and insights in the cosmological theories (especially String Theory) of modern physics. In that sense, this study provides a more demanding, Islamic parallel to such recent popular works such as F. Capra's *Tao of Physics*.

While the prolific Andalusian Sufi writer Ibn 'Arabî (1165–1240) is most widely known today as a mystic and spiritual teacher, his voluminous writings – and particularly his immense magnum opus, the *Meccan Illuminations*, which is the primary source for this study – constantly refer to the insights, theories, and cosmological schemas of earlier Muslim philosophers and scientists, such as Avicenna and the popular spiritual treatises of the 'Brethren of Purity' (*Ikhwân al-Safâ'*). For that reason, this book begins with a helpful survey of the standard theories of cosmology and time found in earlier Hellenistic thinkers, which were largely taken over into the succeeding traditions of Islamic philosophy and science. However, the most creative and unfamiliar aspects of Ibn 'Arabî's cosmological ideas – especially his distinctive conception of the ever-renewed, ongoing and instantaneous nature of the cosmic process of creation (*tajdîd al-khalq*) – are carefully woven together from what have always been profoundly mysterious, problematic, and complexly interwoven symbolic formulations in the Qur'an. Thus the main focus and novel scholarly contribution of the central chapters of this volume lie in the author's careful unfolding and clarification of the intended meanings and references of this dense Qur'anic cosmological symbolism of time and creation, as that multi-dimensional world-view is systematically expounded in elaborate accounts scattered throughout several of Ibn 'Arabî's major works. Every reader who engages with this demanding discussion will come away, at the very least, with a heightened appreciation of the symbolic richness and challenging intellectual dilemmas posed by this unduly neglected – yet arguably quite central and unavoidable – dimension of the Qur'an and its metaphysical teachings.

In the penultimate chapter of this study, before taking up possible analogies to Ibn 'Arabî's ideas in modern physics, the author turns to the language of ontology and to a subject – the paradoxical relations of the divine One and the many – far more familiar to students of Ibn 'Arabî, or of comparable forms of thought in earlier Neoplatonism and the metaphysics of other world religions. Despite the initial unfamiliarity (for non-specialists) of some of Ibn 'Arabî's Qur'anic symbolism and technical terminology here, his approach to conceiving and intellectually explaining the mysterious relationship between the divine Source and its infinite manifestations clearly mirrors Plato's classical dialectical enumeration of the alternative ontological hypothesis outlined in his *Parmenides*. Today, of course, no one is used to thinking of those recurrent metaphysical problems in terms of the theological language of creation. But by this point Dr Haj Yousef has outlined just how Ibn 'Arabî, by carefully elaborating the complex literal indications of the Qur'an itself, is able to illuminate both the temporal and the ontological dimensions of the divine cosmogonic Origination of all things.

The fascinating 'phenomenology' of the human psychological and experiential dimensions of this cosmic creative process, we might add, is also the subject of even more fascinating discussions in Ibn 'Arabî and later Islamic philosophers (as well as earlier Sufis and mystical thinkers). But the elaboration of that closely related topic would require another, equally wide-ranging and original study. So the author has prudently set that related issue aside while focusing on those dimensions of ontology and time most directly connected with the analogous approaches of modern theoretical physics that he outlines in his concluding, more speculative chapter.

This constantly challenging and thought-provoking study is clearly the fruit of years of research on one of the most difficult subjects to be found in the writings of one of Islam's most seminal, creative, inspired, and notoriously difficult thinkers. So even those who may find Ibn 'Arabî's language and speculations difficult to follow will surely come away from their reading with a heightened appreciation of the relative poverty, thoughtlessness and lack of sophistication in today's dominant public discourse about religion and science, and in our prevailing ways of conceiving and approaching these fundamental human issues of cosmology, ontology and theology.

James W. Morris, Boston College

Preface

Ibn 'Arabî is one of the most prominent figures in Islamic history, especially in relation to Sufism and Islamic philosophy and theology. In this book, we want to explore his cosmology and in particular his view of time in that cosmological context, comparing his approaches to the relevant conclusions and principles of modern physics whenever possible. We shall see that Ibn 'Arabî had a unique and comprehensive view of time which has never been discussed by any other philosopher or scientist, before or even after Ibn 'Arabî. In the final two chapters, in which we shall discuss some of the ways his novel view of time and cosmology may be used to build a complete model of the cosmos that may deepen and extend our understanding of the world, while potentially solving some of the drawbacks and paradoxes in the current cosmological models of modern physics.

As we discuss in the opening chapter, there is no doubt that time is one of the most important issues in physics, cosmology, philosophy and theology, and hundreds of books and articles have been published in these fields. However, none of these studies had fully developed Ibn 'Arabî's unique view of time in its cosmological dimensions, although his conception of time is indeed central to understanding, for example, his controversial theory of the 'oneness of being'. One possible reason for this relative neglect is the difficult symbolic language he usually used. Also, he did not discuss this subject at length in any single place in his extant works – not even in chapters 59, 291 and 390 of the *Futûhât* whose titles relate directly to time – so we must piece together his overall cosmological understanding of time from his scattered treatments in many works and different contexts within his magnum opus, the *Futûhât*, and other books. Therefore this book may be considered the first comprehensive attempt to set forth all the relevant dimensions of time in Ibn 'Arabî's wider cosmology and cosmogony.

In Chapter 2, after briefly discussing the different physical theories and models of the cosmos, we start by describing Ibn 'Arabî's cosmos in some detail. Then we also give an extensive review of the different philosophical views of time and its properties from the philosophical and scientific point of view to show the importance of the subject and relate it to Ibn 'Arabî's model. Then, in Chapter 2, we begin to introduce Ibn 'Arabî's general concepts of time and 'days', which are then developed in greater detail in each of the succeeding chapters.

To start with, Ibn 'Arabî considers time to be a product of our human 'imagination', without any real, separately existing entity. Nevertheless, he still considers it to be one of the four main constituents of existence. We need this imagined conception of 'time' to chronologically arrange events and what for us are the practically defining motions of the celestial orbs and other physical objects, but for Ibn 'Arabî real existence is attributable only to the actually existing thing that moves, not to motion nor to time (or space) in which this motion is observed. Thus Ibn 'Arabî distinguishes between two kinds of time – natural and para-natural – and he explains that they both originate from the two forces of the soul: the active force and the intellective force, respectively. Then he explains that this imaginary time is cyclical, circular, relative, discrete and inhomogeneous. Ibn 'Arabî also gives a precise definition – drawing on the specific usage of the Qur'an and earlier Arab conceptions of time – of the day, daytime and night, showing how these definitions are related to the relative motions of the celestial orbs (including the Earth), where every orb has its own 'day', and those days are normally measured by our normal observable day that we count on the Earth.

In Chapter 3 (and also in Chapter 6), we explain the central significance, in Ibn 'Arabî's notion of time and cosmology, of the divine 'Week' of creation, and we begin to develop some of its interesting consequences. To begin with, Ibn 'Arabî considers the cosmic, divine Week, rather than the day or any other time unit, as the main primitive time cycle. Thus he explains how the world is created in seven (cosmic, divine) 'Days', what happens on each Day, and the underlying ontological relation between the Week's Days of creation and the seven fundamental divine Names of Allah. Ibn 'Arabî also shows that all the Days of this cosmic Week, including the last Day (Saturday), all actually occur in Saturday, the 'Day of eternity'. This complex understanding of the ever-renewed divine creation in fact underlies his conception of the genuine unification of space and time, where the world is created 'in six Days' (from Sunday to Friday) as space, and then is displayed or manifested on Saturday in the process that we perceive as time. However, we perceive this complicated process of creation in Six Days and the subsequent appearance of the world on the seventh Day, we perceive all this only as one single moment of our normal time. In fact, on the basis of Qur'anic indications and the corresponding experiential confirmations of the mystical 'knowers' (*'urafâ'*) (later explained in Chapter 5), Ibn 'Arabî insists that the entire created world ceases to exist immediately and intrinsically right after its creation, and that then it is re-created again and again. For him, this process of divine re-creation happens gradually (in series), not at once: i.e. it always takes six divine 'Days' to be prepared and the last Day to manifest. However, we – the creatures – do not witness this re-creation in six Days, since we witness the created world only in the seventh Day (Saturday, which he calls 'the Day of eternity'). So the creation of the world in six Days actually happens every moment, perpetually and recurrently. Therefore, those first six divine Days are actually the creative origin of space and not time, which is only the seventh Day. In this novel conception, for the first time in history, the 'Week', as the basic unit of space-time, will have a specific and quite essential meaning in physics and cosmology.

Even more important in Ibn 'Arabî's conception of time, however, is his understanding of the 'Day' of creation as a minimum indivisible Day, a kind of 'instant of time' (*al-zaman al-fard*) that also includes (since it includes all of creation) the instants of that normal day itself which we live in and divide into hours, minutes, seconds and so on. In order to explain this initially paradoxical notion, Ibn 'Arabî introduces – again initially mysterious Qur'anic on the basis of indications – the different nature and roles of three very different kinds of compounded days (the 'circulated' days, the 'taken-out' days and the 'intertwined' days), which highlight the fact that the actual flow of time is not as uniform and smooth as we feel and imagine. The key concept underlying these complex developments is that Ibn 'Arabî emphasizes, following the Qur'an, that only one creative 'event' should be happening on every Day (of the actual cosmic, divine Days of creation), and not the many different (temporal and spatial) events that we observe. To reconcile this apparent contradiction between the unitary Act (and 'instant') of Creation and the apparent phenomena of spatial and temporal multiplicity, he reconstructs the normal, observable days that we actually perceive in a special manner that is complexly grounded in the different divine 'Days' of the actual flow of time. We shall explain his complex conception of these very different types of days in detail in Chapter 4.

The principle of perpetual re-creation, one of the more famous elements of Ibn 'Arabî's cosmology and cosmogony, is fully explained in Chapter 5, where we also take up the related question of Ibn 'Arabî's controversial theory of the 'oneness of being'. This theory can be easily understood once we have grasped his underlying conception of the eternally renewed creation in time. This comprehensive cosmological vision, when added to his understanding of the actual flow of time based on the three kinds of days described in Chapter 4, can be used to build a new unique model of the cosmos. This cosmological model, which we shall call 'the Single Monad model', is explained in Chapter 6. We shall see in this chapter that, according to this distinctive perspective on creation, the manifest world works exactly like a super-computer which – despite its tremendous speed – can do only one job at a time, where the display on the computer monitor is analogous to the manifest world: though we appear to see a complex, continually changing picture on the screen, that complex image is actually built one pixel at a time by one single electron-beam. This particular illustration helps us to grasp the actual functioning of Ibn 'Arabî's central conception of the ultimate oneness of being, despite the undeniable visible multiplicity of the world.

Finally, Chapter 7 is devoted to discussing some of the implications of the Single Monad model for various related principles of modern physics and cosmology, including the possibilities of testing such a cosmological model. We shall discuss in particular some of the known time-related paradoxes in current models of physics and cosmology, and how they may be resolved according to this novel view. It can be fairly said that Ibn 'Arabî's view of time and the cosmos is a fruitful concept that potentially bridges the gap between traditional theological metaphysical views of the world and the contemporary scientific views that are based on experimental procedures and logic. In addition to explaining the 'oneness of being' and 'creation in six Days', other important results of Ibn 'Arabî's unique concept of time include the ways it helps to resolve the famous EPR paradox, thus potentially reconciling the two great theories of Quantum Mechanics and Relativity in modern physics, how it offers a new understanding of the historical Zeno's paradoxes, and how it potentially explains the reason behind quantization, how quantities are either discrete or continuous.

Acknowledgement

I owe the full appreciation and acknowledgement to Prof. James W. Morris, without whom this work would not have been completed. His precious comments and discussions were the rivers that produced this lake. I also would like to thank Prof. Ian R. Netton, in whose series this book appears, for his encouragement and continuous support. I would also like to thank all my friends and colleagues with whom I had great and important discussions and those who sent me various feedback and criticism. I just would like to mention here some of them: Prof. Radi al-Jishi, Prof. Ahmed Barbour, Dr Ahmed S. Achter, Mr Ahmad T. Summakiyya, and Mr Jum'a al-Ahmed.

Abbreviations

This book is mainly based on Ibn 'Arabî's major comprehensive work *Al-Futûhât Al-Makkiyya*. Because we refer to this book often, we shall use a short reference style: each quotation from the *Futûhât* is followed by a reference in brackets: [X.000.00] which means: [volume.page.line]. When the line number is omitted – in rare cases – this means that the reference is in the entire page; in a few cases a range of a few pages is indicated as [II.229–231]. We have mainly used the standard edition of the *Futûhât* issued by many publishers based on a reproduction of the old edition of Bulaq 1329/1911 which comprises four volumes each about 600–700 pages of 35 lines; the page size is 20 by 27 cm (see the Bibliography). You should notice, however, that newer reset editions may be different although some of them are also in four volumes.

In addition to the *Futûhât*, we use short form of references to many other books by Ibn 'Arabî (for example *Ayyâm Al-Sha'n* and *Al-Masâ'il*) and other related famous works (such as William Chittick's two important works: *The Sufi Path of Knowledge* [*SPK*] and *The Self-disclosure of God* [*SDG*]) as explained in the Bibliography. In most of these cases we put the short form of the reference followed by the page number, with only two exceptions: for Osman Yahya's classification of Ibn 'Arabî's books and for Ibn 'Arabî's *Al-Masâ'il*, where we refer to the listed item number instead of the page number.

All references to the Holy Qur'an are given also in the text and not in notes; after each verse quoted or meaning indicated in the text we add a reference (xx: y) where xx refers to the number of the Sura (chapter) and yy is the number of the Aya (verse). For the prophetic narrations we have mainly used *Kanz al-'Ummâl* (*The Treasure of the Workers*) which includes most of the known *hadith* books, indexed and numbered according to the subject. We give the number of the hadith as [*Kanz*: 0000]. The table below shows the abbreviated references in the book.

Abbreviation	**Original book**
Kanz	Al-Muttaqî Al-Hindî (1989) *Kanz Al-'Ummâl* (*The Treasure of the Workers*), Beirut: Mu'assasat Al-Risâla
Futûhât	Ibn 'Arabî (n.d.) *Al-Futûhât Al-Makkîyya*, Vols 1–4, Beirut: n.p. (see above)
Fusûs	Ibn 'Arabî (1946) *Fusûs Al-Hikam*, Cairo: n.p.; critical edition by Abu 'Alâ 'Affîfî
Dîwân	Ibn 'Arabî (1996) *Dîwân Ibn 'Arabî*, Beirut: Dâr Al-Kutub Al-'Ilmiyya, ed. Ahmad Hasan Basaj
Ayyâm Al-Sha'n	Treatise number 5 in Ibn 'Arabî (n.d.) *Rasâ'il Ibn 'Arabî*, Beirut: Dâr Ihyâ' Al-Turath Al-'Arabi; reprint of Hyderabad ed.
'Uqlat Al-Mustawfiz	Ibn 'Arabî (1919) *'Uqlat Al-Mustawfiz*, including *Kitâb Inshâ' Al-Dawâ'ir* and *Kitâb Al-Tadbîrât Al-Ilâhiyya fî Islâh Al-Mamlaka Al-Insâniyya*, Leiden: Brill, in *Kleinere Schriften des Ibn Al-'Arabî*, ed. H. S. Nyberg
Inshâ' Al-Dawâ'ir	Ibn 'Arabî, *Kitâb Inshâ' Al-Dawâ'ir*, in *Kleinere Schriften des Ibn Al-'Arabî*, ed. H. S. Nyberg
Al-Tadbîrât Al-Ilâhiyya	Ibn 'Arabî, *Al-Tadbîrât Al-Ilâhiyya fî Islâh Al-Mamlaka Al-Insâniyya*, in *Kleinere Schriften des Ibn Al-'Arabî*, ed. H. S. Nyberg
Al-Durrat Al-Baydâ'	Ibn 'Arabî (2002–04) *Risâlat Al-Durrat Al-Baydâ'* in *Rasâ'il Ibn 'Arabî*, Beirut: Mu'assasat Al-Intishâr Al-'Arabî, ed. Sa'îd 'Abd Al-Fattâh, vol. II: 131–145
Al-Tanazzulât Al-Layliyya	Ibn 'Arabî (2000) *Al-Tanazzulât Al-Layliyya fî Al-Ahkâm Al-Ilâhiyya* in *Majmû'at Rasâ'il Ibn 'Arabî*, Beirut: Dâr Al-Mahajjat Al-Baydâ, vol. II: 5–66
Al-Tanazzulât Al-Mawsiliyya	Ibn 'Arabî (2000) *Al-Tanazzulât Al-Mawsiliyya fî Asrâr Al-Taharât Wa-l-Salawât Wa-l-Ayyâm Al-Asliyya* in *Majmû'at Rasâ'il Ibn 'Arabî*, Beirut: Dâr Al-Mahajjat Al-Baydâ, vol. II: 67–314
Kitâb Al-Azal	Book 11 in volume I of *Rasâ'il Ibn 'Arabî*, Beirut: Dâr Ihyâ' Al-Turâth Al-'Arabî, n.d.), which is a photographic reprint of the same famous collection published by the Dâ'irat Al-Ma'ârif Al-'Uthmâniyya (Hyderabad, 1948)
Al-Mu'jam Al-Sûfi	Al-Hakîm, S. (1981) *Al-Mu'jam Al-Sûfî: Al-Hikma fî Hudûd Al-Kalima* (*The Sufi Dictionary: The Wisdom in the Word*), Beirut: Dandarah

SPK	Chittick, W. C. (1989) *The Sufi Path of Knowledge: Ibn Al-'Arabi's Metaphysics of Imagination*, Albany, NY: SUNY Press
SDG	Chittick, W. C. (1998) *The Self-disclosure of God: Principles of Ibn Al-Arabi's Cosmology*, Albany, NY: SUNY Press
OY	Yahya, O. (1964), *Histoire et classification de l'œuvre d'Ibn Arabi*, 2 vols, Damascus: Institut Français
EP	*The Encyclopedia of Philosophy*, ed. by Paul Edwards (1967), 8 vols, New York: Macmillan Publishing
EI^2	Bianquis, Th., Bosworth, C. E., Vandonzel E. J. and Heinrichs, W. P. (eds) (1987) *Encyclopaedia of Islam*, 9 vols, Leiden: Brill

1 Cosmology and time

Cosmology is the science that studies the universe, the cosmos. *Cosmos* is a word used in earlier Greek metaphysical thought that means ‘harmony’ or ‘order’, as opposed to *chaos*. In one Greek theory of creation, *chaos* is the formless matter from which the *cosmos*, or harmonious order, was created (*EP*: ‘Cosmology’, II: 237–244; ‘Chaos and Cosmos’, II: 80–81). And time is one of the most fundamental issues in philosophy and cosmology, since the whole of existence is nothing but consecutive series of events in time. Everybody feels time, but most people do not question it because it is commonly experienced every day in many things and is so familiar. However, it is far more difficult to understand the philosophical nature of time and its characteristics.

Throughout the history of philosophy, many opposing views have emerged to discuss and describe the different aspects of time, and some novel hypotheses have eventually emerged in modern cosmology. However, it is still the dream of every physicist to unveil the reality of time, especially since all modern theories have come to the conclusion that time is the key.

1.1 A brief overview of early cosmological models

Beginning in the twelfth century, Arab scholars, scribes and various translators gradually introduced Europe to the science of astronomy as it had developed in Islamic civilization, based on earlier Hellenistic models (primarily Ptolemy and Aristotle). But once the Catholic Church had decided to adopt the Ptolemaic or Aristotelian geocentric[1] cosmological model as a theological principle, it considered scientists who criticized this model as heretics. Therefore, the Polish scientist Nicolai Copernicus (AD 1473–1544) circulated his heliocentric model anonymously, and his book *De Revolutionibus Orbium Caelestium* (‘On the Revolutions of the Heavenly Orbs’), was not published until 1543, just one year before his death. In this model, Copernicus postulated that the Sun and the stars are stationary and the Earth and the planets circulated around the Sun in circular orbits.[2]

It was not until 1609, when Galileo invented the telescope, that Aristotle’s and Ptolemy’s geocentric model of the universe was completely discarded by knowledgeable researchers, and replaced by the heliocentric model (Drake 1990:

145–163). At around the same date (1609–19), the scientist Johannes Kepler formulated three mathematical statements that accurately described the revolution of the planets around the Sun. In 1687, in his major book *Philosophiae Naturalis Principia Mathematica*, Isaac Newton provided his famous theory of gravity, which supported the Copernican model and explained how bodies more generally move in space and time (Hall 1992: 202).

Newton's mechanics were good enough to be applied to the solar system, but as a cosmological theory it was completely false in so far as he still considered, like Aristotle, the stars to be fixed and the universe outside the solar system to be static. Although a dynamic universe could easily be predicted according to Newton's theory of gravity, the belief in the Aristotelian static universe was so deep and strong that it persisted for some three centuries after Newton (Seeds 1990: 86–107).

In 1718, Edmund Halley compared the positions of stars recorded by the Babylonians and other ancient astronomers with the latest observations and realized that the positions of some of the stars were not the same as they had been thousands of years earlier. Some of the stars were in fact slightly displaced from the rest by a small but noticeable amount. This is called 'proper motion', which is the apparent motion of the star (perpendicular to the line of sight) in relation to the background stars that are very far away. In 1783, William Herschel discovered the solar motion, the Sun's motion relative to the stars in its galactic neighbourhood. Herschel also showed that the Sun and other stars are arranged like the 'grains of abrasive in a grindstone' (Ferguson 1999: 162–165), which is now called the Milky Way galaxy. More than a century later, in 1924, Hubble was able to measure distances to some stars (based on the 'redshift'),[3] and he showed that some bright dots that we see in the sky are actually other galaxies like ours, although they look so small because they are very far away (Hartmann 1990: 373–375).

The Aristotelian theory of a static universe (i.e. of all the stars) had to be reviewed after Hubble's discovery of the redshift of light coming from all distant stars, which indicated that everything in the universe is actually moving; just as Ibn 'Arabî had said many centuries before. In his bestselling book of the 1980s, Stephen Hawking says:

> Even Einstein, when he formulated the general theory of Relativity in 1915, was so sure that the universe had to be static that he modified his theory to make this possible, introducing a so-called cosmological constant into his equations.
>
> (Hawking 1998: 42)

This of course was soon proved to be wrong, and everybody now knows that the cosmos is in continuous motion. Einstein himself later considered this to be one of his greatest mistakes. Ibn 'Arabî, however, declared plainly that the stars cannot be fixed at all, and he even gave numbers and units to the speed of their proper motion [III.548.28, II.441.33], which are consistent with the latest accurate measurements.

After these developments, and with the advent of new technologies employed in making even more accurate observations, in addition to accelerated research in physics and astronomy, a whole new view of the cosmos finally replaced the ancient short-sighted ones. However, we cannot ever claim that all the questions have been answered and that we have drawn a fully correct picture of the cosmos. On the contrary, new sets of even more profound questions are still a riddle, such as dark matter and the Einstein-Podolsky-Rosen (EPR) paradox (see section 6.6).

Along with the vast amount of data collected by telescopes and space shuttles in recent decades, many new theories have arisen to try to explain those observations. The mere concepts of 'time' and 'space' were in focus especially after the strange and courageous ideas of Einstein about relative and curved space-time were proved by Eddington through the observation of the total eclipse of the Sun in 1918 in South Africa. Since then, other theories including Quantum Mechanics, the Field Theory, the Superstrings Theory, and Quantum Gravity Theory, have tried to discover and describe the actual relation between material objects and energy on one hand, and between space and time on the other hand. Yet no fully convincing view has ever been achieved.

1.2 Modern cosmology

Since Copernicus' time, our view of the cosmos has grown both larger and more accurate. It is not our purpose here to explain the modern complicated theories of cosmology, but simply to summarize the present picture of the cosmos as seen by scientists. Our modern picture of the cosmos dates back only to 1924, when Edwin Hubble showed that our galaxy is not the only one in space; many of the faint spots of light that we see in the sky are in fact other galaxies as large as our own, but we see them so small only because they are extremely far deep in space.

Owing to the force of gravity, everything in the sky is moving or orbiting around some point in space. The Moon orbits around the Earth, and the Earth and other planets orbit around the Sun, which also orbits – along with other hundreds of thousands of millions of stars – around the centre of the Milky Way galaxy, which is in turn one of thousands of millions of galaxies all flying through the vast distances of space.

In order to give a clear spatial view of this immense universe, it is better to use big units of distance instead of using big numbers. The best accepted units of distance in cosmology are the 'light year' (9,500,000,000,000,000 metres), which is the distance travelled by light in one year, and the 'parsec', which equals 3.26 light years. Light travelling at 300,000 km/sec can go seven times around the Earth (which has a circumference of approximately 44,000 km) in one second, but it takes 8.33 minutes to reach us from the Sun (150,000,000 km). Proxima Centauri, the nearest star to us apart from the Sun, is about 4.24 light years away. Our galaxy, like most other galaxies, is a collection of about 200 billion stars plus thousands of clusters and nebulae that form together a disk of more than 100,000 light years in diameter, and that is about 15,000 light years thick. The nearest galaxy to us lies in the Andromeda constellation, and it is about 2.9 million light years away. Then galaxies are grouped in somewhat irregular clusters that greatly differ in size from millions to hundreds of millions of light years. The most distant objects discovered so far are about 13 billion light years away. These numbers are simply approximate, just to give an idea of where we are (Hartmann 1990: 413).

It is now also well established that everything in the world is moving: nearby stars have proper motion, because they are pulled towards the centre of the galaxy, and galaxies are moving away from us, because the universe is expanding. On the other hand, and despite these various motions, the universe does not have a centre or edges. It is hard to imagine, but the universe is contained or curved around itself so that if you fly straight in one direction and keep moving in a straight line you will one day, if you live long enough, come back from the opposite direction to the same point (supposing no gravitational fluctuations), just as it would happen to a person travelling around the Earth.

The stars that we see in the sky are, just like our Sun, huge nuclear fusion reactors that are constantly converting hydrogen into heavier elements and hence producing heat and light. But not all stars are the same: some are big and some are small; some are young and some are old; some are bright and some are faint. Also, many stars are dying and many others are born all the time in a process of very complicated evolution (Seeds 1990: 134–281).

So how is all this explained according to the new cosmological theories? We can not discuss here all the different theories in physics and cosmology, but we want to make a quick summary of the basic principles of the different models of the cosmos so that we can understand the potential importance of the 'Single Monad model' which we are going to propose in the last chapter of this book, based on Ibn 'Arabî's unique understanding of time and his famous theory of the oneness of being.

1.3 Summary of modern theories of cosmology

After the amazing discoveries and the enormous amount of data obtained by telescopes and space shuttles, and with the success of the theories of Relativity and Quantum Mechanics, scientists tried to build new cosmological models to explain the structure and origin of the universe on the basis of the new information. We shall give here a very short summary of the major theories of cosmology that have developed recently.

Scientists up to the beginning of the twentieth century believed in a stationary universe outside the solar system, but this was soon proved to be wrong. Actually the same theory that Einstein first tried to make fit a steady universe and fixed stars later proved that the universe is expanding. This implied that the universe had started at one moment, about 15 billion years ago, from a very small point, but with very high density, and then it expanded to its present state. This was called the 'Big Bang', and many cosmological models were developed on the basis of this view (Narlikar 1993: ch. 2, ch. 5).

The 'Steady State' theory tried to explain the expansion of the universe by supposing a continuous creation of matter that filled the space produced by the expansion, but the discovery of cosmic microwave background radiation in 1965 by Penzias and Wilson caused the Steady State model to be completely discarded. The background radiation was interpreted as the faint afterglow of the intense radiation of a 'Hot Big Bang', which had been predicted by Alpher and Hermann back in 1949, although some people also attribute it to Gamov back in 1946 (Dolgov *et al.* 1990: 11).

The problem with the background radiation was that all measurements showed it to be very uniform in all directions. This isotropy of the background radiation was a riddle because with homogeneity no stars or galaxies could be produced (Tayler 1993: 194). It was only in 1992 that NASA's Cosmic Background Explorer satellite (COBE) detected the first anisotropies in this background radiation: one part in a hundred thousand, which may indicate the seeds from which galaxies formed (Schewe and Stein 1992).

The Big Bang model was very good in explaining many of the observations, yet on the other hand there were many contradictions (Linde 1990: 4). Many of these theoretical contradictions were resolved by the 'inflationary scenario' devised by Alan Guth in 1979. Guth looked at a very early stage in the development of the universe from about 10^{-32} to 10^{-43} of a second after the initial creation. During this period matter was in very highly excited states, causing the most extreme conditions of high density and pressure which made the cosmos expand exponentially, filling the universe with an intense dense fire of particles and photons (Linde 1990: 42).

In classical (Newtonian) mechanics, one could predict the behaviour of a system if one exactly knew its initial state. But in Quantum Mechanics, we can only calculate the probability of how the system will evolve (White 1966: 29). In either case, however, the main problem in cosmology is to determine the initial state that the laws should be applied to. One successful approach to get round this problem is to work backwards by using the observed properties of the universe to deduce what it was like in an earlier state.

The problem with the inflationary theory is that, in order for inflation to have occurred, the universe must have been formed containing some matter in a highly excited state, but the next question is why this matter was in such an excited state. To overcome this, some scientists tried to apply Quantum Mechanics to the whole universe, and the result was the theory of Quantum Cosmology.[4] This may sound absurd, because typically large systems (such as the universe) obey classical, not quantum, laws. Einstein's theory of General Relativity is a classical theory that accurately describes the evolution of the universe from the first fraction of a second of its existence up to now. However it is known that General Relativity is inconsistent with the principles of Quantum Theory, and is therefore not an appropriate description of the physical processes that occur at very small length scales or over very short times. To describe such processes we require the theory of Quantum Gravity.

In non-gravitational physics, the approach to Quantum Theory that has proved most successful involves mathematical objects known as 'Path Integrals' that were introduced by the Nobel Prize winner Richard Feynman. In the Path Integral approach, the probability that a system in an initial state *A* will evolve to a final state *B* is given by adding up a contribution from every possible history of the system that starts in *A* and ends in *B*. For large systems, contributions from similar histories cancel each other in the sum and only one history is important. This history is the history that classical physics would predict. At any moment, the universe is described by the geometry of the three spatial dimensions as well as by any matter fields that may be present. Given this data, one can in principle use the Path Integral to calculate the probability of evolving to any other prescribed state at a later time. However, this still requires knowledge of the initial state.

Quantum Cosmology is a possible solution to this problem. In 1983, Stephen Hawking and James Hartle developed a theory of Quantum Cosmology which has become known as the 'No Boundary Proposal'. In practice, calculating probabilities in Quantum Cosmology using the full Path Integral is formidably difficult and an approximation has to be used. This is known as the 'semi-classical approximation', because its validity lies somewhere between that of classical and quantum physics. In the semi-classical approximation, one argues that most of the four-dimensional (space-time) geometries occurring in the Path Integral will give very small contributions to the Path Integral and hence these can be neglected, so we can deal only with three dimensions (space). The Path Integral can be calculated by considering just a few geometries that give a particularly large contribution. These are known as 'Instantons' (from 'the instant', because it aims at omitting time, so it is like a snapshot that takes into account only the three co-ordinates of space), which describe the spontaneous appearance of a universe from literally nothing. In this way we do not have to think about the cosmos as something that takes place inside some bigger space-time arena. Once the universe exists, Quantum Cosmology can be approximated by General Relativity, so time appears.

Research in these areas is still ongoing, but one of the many outstanding problems in trying to construct a Quantum Field theory of gravitation concerns the appropriate interpretation of quantum states for configurations that make no overt reference to 'time'. We shall see by the end of this book that Ibn 'Arabî's understanding of time could be a key to eliminating these peculiarities, because he simply views the world as an eternal existence that is perpetually being re-created. He also unified space and time in a manner that has apparently never been thought of before or since.

1.4 Preliminary outline of Ibn 'Arabî's cosmology

Ibn 'Arabî (AH 560–638/AD 1165–1240) was a great Sufi thinker of the Middle Ages and one of the most influential authors in Islamic history, whose writings have deeply influenced Islamic civilization for centuries, and have more recently attracted wide interest in the West. The full name of Ibn al-'Arabî (more commonly referred to in English without the definite article) is Abû 'Abd Allâh Muhammad Ibn al-'Arabî al-Hâtimî al-Tâ'î. He was born in Murcia (in eastern Andalusia), into a very pious and cultured family. When he was seven they moved to Seville, and at the age of 16 he 'entered on the path' (of Sufism). Then he travelled throughout and between Andalusia and Morocco for some years before a vision compelled him to go to the East. In 1201 he travelled to Cairo, al-Quds (Jerusalem), and finally to Mecca for pilgrimage. His many works eventually brought him fame, and sometimes notoriety, so that he was eventually sought out by Seljuq and Ayyubid princes and accompanied by a group of disciples. Later on he came to be popularly called *Muhyî al-Dîn* ('Reviver of Religion') and *al-Shaykh al-Akbar* ('the Greatest Master'). He continued travelling throughout the Middle East until he settled in 1224 in Damascus, where he remained until his death in 1240.[5]

Ibn 'Arabî's two most famous and influential works are *Al-Futûhât Al-Makkiyya* (*The Meccan Illuminations*), an encyclopaedic discussion of Islamic wisdom (Nasr 1964: 92–98), and the shorter *Fusûs al-Hikam* (*The Bezels of Wisdom*), which comprises chapters named after prophets who characterize different spiritual types. But Ibn 'Arabî also wrote many other lesser known works, many of them now available in print, such as the *Kitâb Al-Tajalliyyât*, *Tarjumân Al-Ashwâq*, *Mashâhid Al-Asrâr Al-Qudsiyya*, *Mawâqi' Al-Nujûm*, *'Uqlat Al-Mustawfiz*, *Inshâ' Al-Dawâ'ir* and *Al-Tadbîrât Al-Ilâhiyya*, in addition to 29 shorter treatises published in the Hyderabad collection commonly known as the *Rasâ'il Ibn 'Arabî*, and many other shorter books and treatises. In one of his treatises, Ibn 'Arabî himself listed 289 titles, which increase to 317 confirmed works when added to other titles he mentioned throughout his various books. More than 850 books have been attributed to him.[6]

Ibn 'Arabî was not an astronomer, and was never interested in astronomy as a science. But as a Sufi and mystical theologian constantly inspired by the cosmological teachings and symbolism developed throughout the Qur'an and in a number of related Hadith (Prophetic sayings), he talks about planets and orbs and their motion as a structure Allah created on His Image (see section 3.2) and relates them to the divine Names. He uses cosmology to refer to the ways we acquire more knowledge of Allah. Apart from a few short treatises where he talks about some astronomical subjects mixed with philosophy and theology, Ibn 'Arabî did not devote any special book to describing the heavens. Nevertheless, in his major book *al-Futûhât al-Makkiyya* (henceforth referred to as 'the *Futûhât*'), for example, we find many paragraphs that can be used to illustrate his profound view of the cosmos.

It can surely be said that Ibn 'Arabî's view of the cosmos is truly challenging, even as compared to the latest modern theories. For example, he clearly declared that the stars are not fixed at all, more than seven centuries before this was scientifically known, and he explained why we do not see their motion. Moreover, he gave numbers to the average velocities of the proper motion of stars as 100 years per arc degree, which is quite consistent with the measurements taken only few decades ago [III.548.28, II.441.33]; indeed he even used exactly the same unit of measurement now being used (Smart 1977: 249) at a time when no such measurements were possible at all. He also explained the observed 'retrograde motion' of some planets and the formation of the planets in the solar system in a similar manner to what is widely accepted today [II.443.24, III.203.21]. But most important in this regard is that his view of the world is heliocentric, similar to what Copernicus suggested many centuries afterwards. He also clearly affirmed that the Earth is spherical, moving and rotating, and he also explained why people do not realize the motion of the Earth around its centre [I.123.17, II.441.33, III.548.21].

Ibn 'Arabî's unique understanding of the process and reality of ongoing creation has been discussed by many scholars in some details. Ibn 'Arabî himself mentioned in particular a number of key cosmological developments in chapter 371 and in the very detailed chapter 198 of the *Futûhât*, as well as in other cosmological books such as *Inshâ' Al-Dawâ'ir*, *Al-Tadbîrât Al-Ilâhiyya* and *'Uqlat Al-Mustawfiz*. William Chittick has devoted an immense volume called *The Self-disclosure of God: Principles of Ibn al-'Arabi's Cosmology* (this will be abbreviated as *SDG*) specifically to Ibn 'Arabî's cosmology and ontology, in addition to some chapters of other books like *The Sufi Path of Knowledge: Ibn al-'Arabi's Metaphysics of Imagination* (this will be abbreviated as *SPK*), and also Henry Corbin discussed some aspects in his pioneering study, now entitled in English *Alone with the Alone. Creative Imagination in the Sufism of Ibn 'Arabî* (Corbin 1969: 184). Here we want to give a very short summary of Ibn 'Arabî's cosmology, in a way somewhat different from the approach followed by Chittick and Corbin. We only want to give a general description of his cosmological views, without too much further analysis and explanation, so that we can concentrate on the central subject of time in the rest of the book. Also we will leave the discussion of the ontological aspects of his cosmology to the following chapters (see in particular section 3.3). Here in the following we shall use the same figures Ibn 'Arabî drew in chapter 371 of the *Futûhât*, and the following broad cosmological account is mainly drawn from that chapter [III.416–448], along with a few paragraphs taken from the long chapter 198 [II.390–478] of the same work.

Ibn 'Arabî's universe comprises both the material and the abstract, spiritual or noetic (*'aqlî*) worlds. He says that the main reason for creating the cosmos is 'Love'. In explaining this underlying principle he often refers to a famous divine saying (the 'Hadith of the Hidden Treasure')[7] which states that Allah 'loved' to be known in order to grant the creatures the privilege of coming to know Him. Thus Allah's love to be known is a Mercy (*rahma*) from Him that He wanted to grant to all His creatures. This Mercy is the first state of the presence of Allah with regard to the world to be created, and hence it formed the abstract place (or 'space') in which creations would appear. Following indications in another Prophetic Hadith, Ibn 'Arabî calls this abstract place *al-'amâ'* ('the Cloud').[8] According to his account, the reality of *al-'amâ'* accepted the forms of the 'Roaming Spirits' (*al-arwâh al-muhayyama*) that Allah created directly, without any intermediaries. This direct creation caused these angelic Spirits to roam in the presence of Allah, knowing nothing but Him. They did not even know about themselves (i.e. they had no self-consciousness). Allah appointed one of these spirits and granted him a special epiphany of divine Knowledge (*tajallî 'ilmî*) that engraved in him all that Allah wants to create in this entire cosmos until the Last Day. The other primal Spirits could not know about that. This initial epiphany caused this Spirit – which is then called the 'Universal Intellect' (*al-'aql al-kullî*) or the 'First Intellect' (*al-'aql al-awwal*) or also, using a central Qur'anic symbol, the 'Higher Pen' (*al-qalam al-a'alâ*) – to become aware both of himself and of the other Spirits, while they did not know about him.

Through this epiphany, the First Intellect saw himself composed of himself and of his further ability to realize or 'intelligize'. He also saw that he has an ontic 'shadow' caused by the Light of that special epiphany, which was realized through the divine Name 'the Light' (*al-nûr*). This shadow is his 'soul', which is called the 'Universal Soul' (*al-nafs al-kulliyya*) or the 'First Soul' (*al-nafs al-ûlâ*), or also the 'Highest/Protected Tablet' (*al-lawh al-a'alâ/al-mahfûz*), in which he is going to write what he knows is going to happen until the Last Day. The entire universe, then, is – to use a central Qur'anic symbolism – the 'letters' and 'words' of Allah that are produced through 'the Breath of the All-Merciful'. We shall see in section 5.8 that the fundamental 'blocks' in the universe are 'strings' or vibrations ('sounds' or 'notes'), which is similar to Ibn 'Arabî's notion of the hierarchy of the 'men of breaths' (*rijâl al-anfâs*). Therefore it is not only a symbolism to say that the entire universe is the 'letters' and 'words' of Allah,[9] and those words are continuously being written by the Highest Pen (the First Intellect) in the Highest Tablet (the Universal Soul). Figure 1.1 shows this Cloud and its contents down to the 'establishing Throne' (*'Arsh al-Istiwâ'*), which is different from the usual cosmological meaning of the divine (normal or usual) 'Throne'. The 'establishing throne' is the throne on which '*Allah established His authority*', alluding to the verse '*ar-Rahmân 'ala al-'arsh istawâ*' (20:5).

According to this account in chapter 371, the universe appeared in the Universal Soul through the Universal Intellect as the result of what Ibn 'Arabî calls an 'abstract (or 'spiritual') marriage' (*nikâh ma'nawî*). This is because everything that happens as a result of to a particular cause is like a 'son' of this cause who is considered its 'father', and its 'mother' is the object where this 'son' appears or happens. Just as we are all (in our outer bodily dimension) the 'children' of Adam and Eve, all other things in the universe can be considered the 'children' of the Universal Intellect and the Universal Soul.

The Universal Soul has two forces mentioned in Figure 1.1: the 'intellective force' (*quwwa 'ilmiyya*) by which it perceives knowledge, and the 'active force' (*quwwa 'amaliyya*) by which it preserves its existence through motion. The first thing the Universal Soul gives rise to, as indicated in the same figure, is twofold: 'the level of Nature'[10] and the 'Chaos' (*al-habâ'*: literally 'the Dust') or 'the Prime Matter' (*al-hayûlâ al-ûlâ*) [I.140.14]. From here on, Ibn 'Arabî uses the symbolic conjugal imagery of the 'wedding' of generative elements and of 'birth' at each successive level of creation or manifestation. Thus the Universal Soul first begets Nature and then Prime Matter or Dust. Then Nature and Dust in

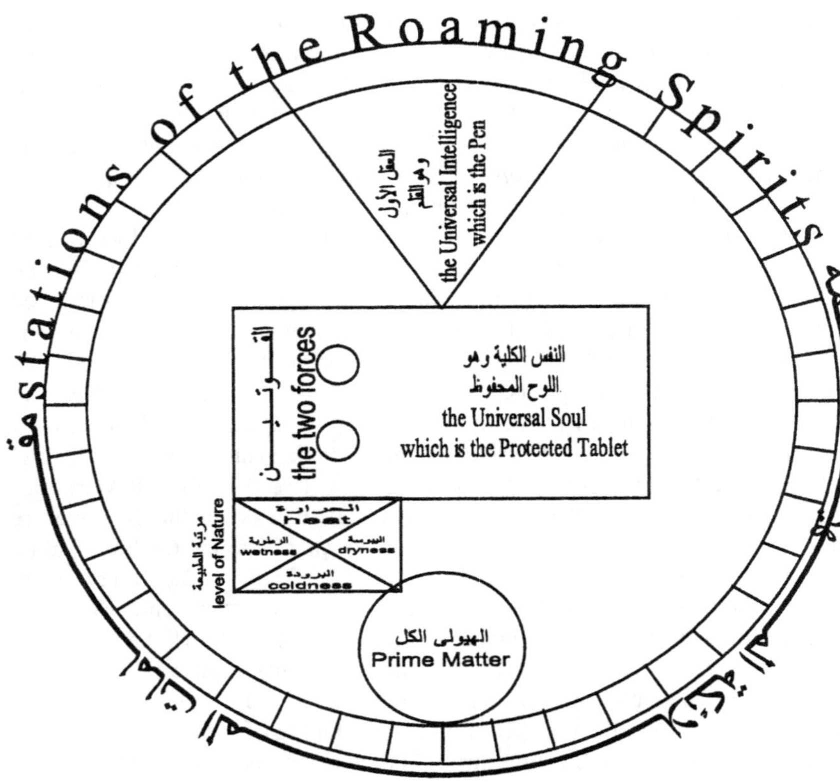

Figure 1.1 'The Cloud' and what it contains, down to the 'establishing Throne'.

Note
This diagram is translated from Ibn 'Arabî's drawing in chapter 371 [III.421].

turn beget their first 'son', which is called the 'Universal Body' (*al-jism al-kull*). This symbolic process of cosmic 'births' continues in a long and defined series of causes and results until it reaches the 'soil' (*turâb*) [I.140.17] which refers to physical matter in general. So the physical world appeared 'after' this Universal Body, while before that all was only spiritual.

As in Figure 1.3, the Universal Body seems to contain everything beneath it including the zodiac (with all the stars and galaxies). Alternatively, we can consider that the physical world is formed by (not 'in') the Universal Body because, like the Universal Intellect and Soul, this Body can be called the First Body because it was the first body to be created. In addition to that, the world both as material and spiritual is formed by the Single Monad through the continuous manifestations of this Monad. If we then consider that the First Body was the first 'elementary particle' to be formed by the Single Monad then the physical world is formed 'by' this First Body. The other possibility is that the Universal Body is some sort of a huge cloud of matter in primary form, which then developed into stars and galaxies, in which case we could say that the physical world is formed 'in' the Universal Body. The first thing which was formed in (or by) the Universal Body was the 'Throne' (*al-'Arsh*) on which Allah established his authority (*istiwâ'*)[11] from His Name 'the All-Merciful' (*al-Rahmân*), which means that all creatures beneath the Throne are to be granted the creative Mercy of their existence from Him. Therefore the first thing that the Highest Pen or First Intellect wrote in the Higher Tablet (the Universal Soul) was this 'Throne' in which the entire creation (the cosmos) is to appear. All this is shown in Figure 1.2.

Inside (or 'beneath') the divine Throne there appeared the 'Pedestal' (*al-Kursî*), whose relative dimensions and plenitude, in comparison to the infinitely vast noetic or spiritual dimension of the 'Throne', Ibn 'Arabî compares here to 'a tiny ring in a vast desert'. And within this 'Pedestal' is the 'Isotropic Orb' (*al-falak al-atlas*), which is shown to contain the sphere of the divisions of the

Figure 1.2 The establishing Throne and what it contains down to the Pedestal.

Note
This diagram is translated from Ibn 'Arabî's drawing in chapter 371 [III.422]. We give the title as it is in the original text, though the diagram shows the Prime Matter and the Universal Body, in addition to the Throne down to the Pedestal.

zodiac (*falak al-burûj*) and the sphere of the stars (*al-falak al-mukawkab*), including beneath them the separate orbs of the five planets, Sun, Moon, and the Earth. All this is shown in Figure 1.3 and Figure 1.4.

The Isotropic Orb or sphere is so called because it contains no stars nor yet any distinguishing feature; it is homogeneous in all directions. The sphere of the zodiac was the first orb to be created inside the Isotropic Orb, and its surface was divided by human convention into the 12 equal parts that are traditionally assigned to the various zodiacal signs. According to the diagram in Figure 1.3 and Ibn ʻArabî's comments on it in chapter 371 of the *Futûhât*, it is evident that he was ware of the large distances between the galaxies because the fixed stars are in our galaxy while the zodiac signs are other galaxies placed very far away. In this fast space Allah created the seven paradisiacal 'Gardens' (*al-jinân*, s. *janna*) named in the Qur'an, with their different states and levels marking the symbolic 'meeting-place' between the purely spiritual realities of the divine

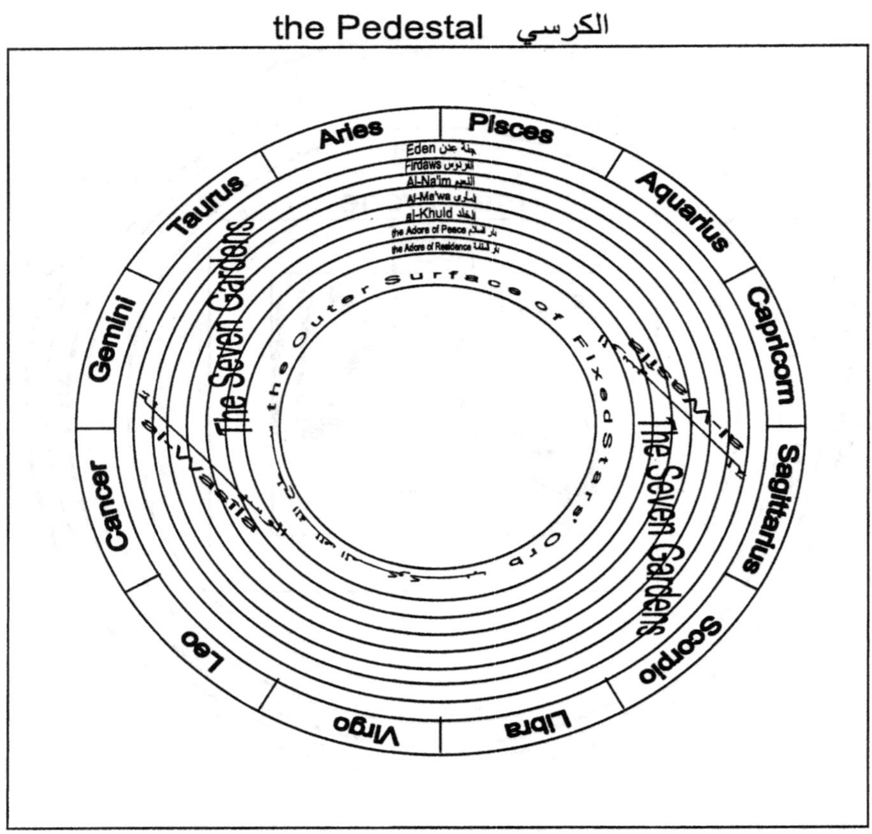

Figure 1.3 The (divine) Pedestal and what it contains down to the constellations.

Note
This diagram is translated from Ibn ʻArabî's drawing in chapter 371 [III.423].

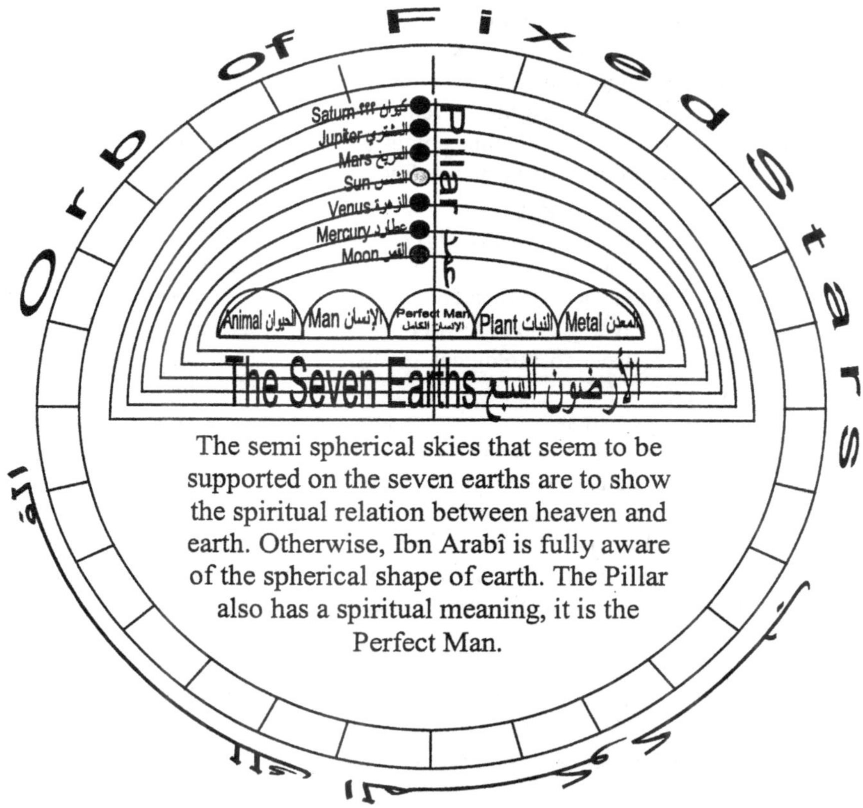

Figure 1.4 The orb of the constellations and what it contains down to the Earth.

Note
This diagram is translated from Ibn 'Arabî's drawing in chapter 371 [III.424].

Throne and the 'sensible' realities in the realm of the Pedestal. The specific names of each of the seven Gardens are taken from related verses in the Qur'an and Hadith, and they are different from the Seven Heavens or Skies (*samawât*) which are, for Ibn 'Arabî, the same seven celestial spheres where the five known planets plus the Moon and the Sun are, as shown in Figure 1.4 and Figure 1.3. The word '*al-Wasîla*' that twice crosses all the seven Gardens (in Figure 1.3) corresponds to 'the highest level in (the highest Garden of) Eden, and it belongs (specifically) to the Messenger (Muhammad) of Allah' [I.319.14, also see I.658.30]. It is also known as '*al-maqâm al-mahmûd*' ('the commendable station'), and it was called '*al-Wasîla*' ('the Intermediary', or 'the Way (of Approach to Allah)') because 'through It Allah may be approached' [II.87.9].

Then beneath the seven Gardens comes the orb of the (apparently) fixed stars, the constellations, and the 'houses' or 'mansions' (*manâzil*) of the Moon.

However, Ibn 'Arabî maintained that those stars are not fixed at all, but that our human time-scale is too short to notice their motion [II.441.33].

The orb of the fixed stars is (also conventionally) divided into 28 constellations or 'houses' through which the Moon appears to pass. Then inside this sphere of the stars, Allah created the 'seven (visible) heavens' (*al-samawât*) and the Earth. And here Ibn 'Arabî again points out that in relation to the divine 'Pedestal' (*Kursî*), the dimensions of our Earth together with the seven visible heavens are like a ring in a vast desert – just as the Pedestal stands in that same relation to the immensity of the divine Throne.

Then Ibn 'Arabî speaks at length (chapter 371 of the *Futûhât*) on the states and levels of the Gardens and Gehenna and other descriptions of the other world (*al-âkhira*). Here, however, we shall restrict ourselves to this very short summary of a few general relevant cosmological points, because of our focus on the concept of time.

First, we should note that Ibn 'Arabî, following normal Arabic usage, also calls the Sun and the Moon 'planets'. But at the same time he clearly distinguishes between the nature of the planets (including the Moon) and the Sun itself, observing that the Sun alone 'is responsible for illuminating all other planets above and below' [II.170.22]. As is normal in Arabic writings (including astronomical ones), he also calls the stars by the same term as 'planets' (s. *kawkab*), yet he also knows that those stars are like the Sun in that they emit their own light [I.217.18].

A first quick reading of Ibn 'Arabî's texts about the world might reveal the same traditional Aristotelian (geocentric) cosmological world-view because, like most other ancient cosmologies (and apparently the Qur'an and Hadith), he talks about 'seven (celestial) heavens' around or above the Earth, each inhabited by a planet (including the Sun and the Moon, as shown in Figure 1.4). But Ibn 'Arabî stresses in many places [III.548.21, I.123.17, II.441.33] that this is only the apparent view for a person who is sitting on the Earth, thus distinguishing between this apparent earthly view and the actual motion of the planets and stars themselves. So, for Ibn 'Arabî, Aristotle's view is a view of the world 'as we see it . . . while in itself it cannot be described like that' [III.548.31]. He stresses the central position of the Sun which he considers to be in the 'heart' (centre) of the seven heavens, and he emphasizes the superiority of the Sun over other planets that are even above it with relation to the Earth: 'So the elevation of this place (the Sun's orb) comes from its being the heart of orbs, so it is a high place for its status and the orbs that are above it in distance with relation to our heads, are still below it in status' [III.441.33]. His actual view of the (local) world is therefore in some sense 'heliocentric', at least in relation to the unique central status or 'rank' (*makâna*) of the Sun.

As for those areas of the sphere of the fixed stars and the visible constellations normally specified by the 12 signs of the zodiac or the 28 houses of the Moon, Ibn 'Arabî considers them as a mere convention, which do not necessarily relate to the actual positions of those particular stars. He says: 'The zodiac (constellations) are approximate positions, and they are houses for the moving planets' [III.37.27]. And for the Moon he says that 'those stars are called "houses" because planets move through them, but otherwise there is no difference between them and other stars that are not houses. . . . They are only assumptions and proportions in this body (of the sky)' [III.436.30].

On the other hand, we cannot strictly separate the material world from the abstract or spiritual world, as they are really overlapped – or rather, all of the material worlds (of the 'Pedestal' and the visible heavens and Earth below) are effectively contained within the immaterial divine 'Throne'. This is why Ibn 'Arabî sometimes mixes the two views: for example he drew a pillar to refer to the Perfect Human Being, whom he considers to be the 'image of the Real' (i.e. of God) in the cosmos, so that without him the cosmos would collapse. He also speaks, following scriptural symbolism, about the seven heavens as being 'supported' on the seven (levels or regions of the) earths. But Ibn 'Arabî does not consider that to be the actual physical picture of things, because he clearly states that the Earth is spherical and that it rotates around its centre: 'but the motion of the Earth is not apparent for us, and its motion is around the middle (centre) because it is a sphere' [I.123.17]. He even nicely explains why we do not feel the motion of the Earth and the cosmos in general (stars). For example he says that people and most other creatures do not feel the motion of the cosmos because it is all moving so the witnessed dimensions don't change, and that is why they imagine that the Earth is stationary around the centre [II.677.21].

1.5 Time in philosophy and science, introduction

Everybody feels time, and most people do not question it because they experience every day and it is so familiar (Fraser 1987: 17–22). But if we want to understand the nature of time we have to answer many basic questions such as:

- What really is time?
- Can we stop it?
- Can we reverse it?
- Is the flow of time universal or is it related to the observer?
- When was the beginning of time, and does it have an end?
- Does time exist objectively, or is it only a construct of our imagination?
- What is the relationship between time and space?
- What is the structure of time?
- Is time continuous or discrete?
- What does the word 'now' or 'moment' mean?
- Why does time move into the past?
- What is the reality of the future?

These and many other similar questions have been the subject of philosophy, physics and cosmology for many centuries, with little progress in finding convincing answers. The question: 'What is time?' is more like the question: 'What is love?', because it is something that everybody can feel it, but no one can give an exact definition of it. If you ask this question to many people, you will certainly get as many answers. St Augustine, in his *Confessions*, asks, 'What is time?' When no one asks him, he knows; when someone asks him, however, he doesn't know (*EP*, 'Time', VIII, 126).

The understanding of time was very important for early people both from the practical view, where they needed to anticipate major events such as floods and harvest time, and from the philosophical view, which is based on sheer curiosity and love of knowledge. Many religions and ancient philosophies, therefore, have tried to answer some questions about time. Some of these religions and philosophies consider time as circular with no beginning or end, some consider it as linear with infinite extension in the past and in the future, and others consider it as imaginary, while real existence is for motion or moving bodies only.

The concept of time is necessary when we ask about the chronological order of things and the duration of events. And because our life is full of events of all types, so time has its signature in all aspects of life. Some examples are: the ageing process in biology, timekeeping in mechanics, the arrow of time and entropy in thermodynamics, and the radically varying psychical time that one feels when waiting for something or in other circumstances. Therefore, in order to understand the reality of time, one needs to explore many closely related fields like physics, biology, psychology and cosmology.

In recent centuries, with the revolutionary new concepts in physics and cosmology in addition to modern technology, increasing accuracy of timekeeping became very important because it is the reference for the extremely complicated motions – of the different parts of a machine for example – that have to work together in coherence. The critical importance of timing events both on Earth and in space was enhanced by precise timekeeping machines like electronic clocks, atomic clocks, and pulsars which are fast-rotating stars that emit short radio pulses at regular intervals with extremely high precision. But despite the new abstract concepts about time like 'time travel' and the 'curvature of time' brought about by the theory of Relativity, our modern concept of time has usually remained quite practical because everything has to be done according to the clock. In fact, the modern theories of physics and cosmology have added more questions and paradoxes about time than they answered (Grünbaum 1971: 195–230).

In general we can detect two major opposing views in the philosophy of time:

1 the rational (realistic) view based on the physical understanding of the world
2 the idealistic (perhaps apparently 'irrational') view based on metaphysics.

Rationalists believe that the mind is the most powerful force of humankind and is able to understand everything in the world, while the irrationalists consider the world, including time, as something beyond the capabilities of our minds. For the idealist, nothing, including time, exists independently of the mind. Therefore the idealist believes that time is a construct of our mind and does not have a separate existence.

1.6 Time in Greek philosophy

Since the age of Homer, the Greek word *chronos* has been used to refer to time. Chronos was a Greek god who feared that his sons would take over his kingdom, so he ate them one after the other – just like time, which brings things into existence and then overtakes them.

We can already detect two clearly opposing views about time in the contrast between Plato's and Aristotle's schools of thought. Plato considers time to be created with the world, while Aristotle believes that the world was created in time, which is an infinite and continuous extension. Plato says: 'Be that as it may, Time came into being together with the Heaven, in order that, as they were brought into being together, so they may be dissolved together, if ever their dissolution should come to pass' (Cornford 1997: 99).

Aristotle, however, believes that Plato's proposition requires a point in time that is the beginning of time and has no time before it. This notion is inconceivable for Aristotle, who adopts Democritus' notion of uncreated time and says: 'If time is eternal motion must also be eternal, because time is a number of motion. Everyone except Plato has asserted the eternity of time. Time cannot have a limit (beginning or end) for such a limit is a moment, and any moment is the beginning of a future time and the end of past time' (Lettinck 1994: 562).

Thus time for Aristotle is a continuum, and it is always associated with motion; as such, it can not have a beginning (Lettinck 1994: 241–259, 361). Plato, on the other hand, considers time as the circular motion of the heavens (Cornford 2004: 103), while Aristotle said that it is not motion, but rather the measure of motion (Lettinck 1994: 351, 382, 390). Aristotle clearly relates rational time and motion, but the problem that arises here is that time is uniform, while some motions are fast and some slow. So we measure motion by time because it is uniform – otherwise it cannot be used as a measure. To overcome this objection, Aristotle takes the motion of the heavenly spheres as a reference, and all other motions, as well as time, are measured according to this motion (Badawî 1965: 90). On the other hand, Aristotle considers time as imaginary because it is either past or future, and both do not exist, while the present is not part of time because it has no extension (Lettinck 1994: 348).

We shall see in Chapter 2 that Ibn 'Arabî shares with Aristotle the idea of a circular endless time and that it is a measure of motion, but he does not consider it as continuum. On the other hand, Ibn 'Arabî agrees with Plato that time is created with the world. In fact Plato was right when he considered time to be created, but Aristotle refused this because he could not conceive of a starting point to the world nor to time. Only after the theory of General Relativity in 1915, which introduced the idea of 'curved time', could we envisage a finite but curved time that has a beginning. By this we could combine Plato's and Aristotle's opposing views. However, Ibn 'Arabî did that seven centuries before, and he also explicitly spoke about curved time a long time before Einstein.

1.7 Time in earlier Islamic philosophy

Muslim philosophers were in general greatly influenced by their Hellenistic predecessors, and therefore tried to apply their theories of time in relation to the related issues raised by the Qur'an and Prophetic Hadith. Many Muslim philosophers prior to Ibn 'Arabî, such as al-Kindî, al-Fârâbî, al-Râzî, al-Ghazâlî, Ibn Rushd (Averroes) and Ibn Sînâ (Avicenna), analysed and criticized or adapted the differing conceptions of time in the schools of Greek philosophy represented by Aristotle, Plato and Plotinus (Badawî 1965).

For example, al-Ghazâlî,[12] in his famous *Refutation of the Philosophers* (*Tahâfut al-Falâsifa*),[13] dealt with most of the standard philosophical and logical arguments regarding time and creation, and his criticism was also thoroughly discussed in Ibn Rushd's famous philosophical rebuttal, the *Tahâfut al-Tahâfut*.[14] However, one of the most influential Muslim philosophers who had many original views about time is Ibn Sînâ, who devoted long chapters in several works to discussing time and related issues according to the views of kalam theology and of previous philosophical schools.[15]

Ibn Sînâ[16] started by summarizing the metaphysical positions of all previous (Islamic and other) philosophers[17] and then criticizing their different ontological views. Although Ibn Sînâ, like Aristotle, closely relates time and motion, he stressed that motion is not the amount of time. He based his argument on the fact that different distances can be cut in the same time, or that the same distance can be done in different times, either owing to the difference in velocity or owing to stops on the way.[18] But ultimately he does define time by motion, though he adds distance to overcome Aristotle's difficulty regarding periodicity (see previous section). He says that time 'is the number of motion when it separates into before and after, not in time but in distance, otherwise it would have to be periodical' (Nasr 1978: 224–226).

On the other hand, although Avicenna doubted the existence of physical time, arguing that time 'exists' in the mind only as a result of to memory and expectation, he also showed that time is also real in the sense that it exists through motion which relates to physical matter; time is real, but it does not have a stand-alone essence, since it exists only through the motion of matter.[19] On the issue of the structure of time, Ibn Sînâ affirms that it is a continuous quantity, since he (like Aristotle) considers time to be the amount of circular motion which is continuous, and thus time is divided only by our mind's illusion into 'moments' or 'instants' (*ânât*).[20]

On the other hand, the proponents of kalam theology, particularly the Ash'arites,[21] on the basis of their atomist view, consider time to be discrete, and they also talked about the re-creation of the world in time. Ibn 'Arabî acknowledges the positive insights in their position, but he also criticized their view as being incomplete (SPK: 97). We shall discuss Ibn 'Arabî's own view in detail in sections 2.8 and 5.6 below.

At the earliest stage of Islamic philosophy, the philosopher and mathematician al-Kindî[22] was generally affected by Aristotle and adopted his view that time is the number of motion.[23] However, arguing from the general Qur'anic principle that Allah is the One Who created the world, he asserted that the material world cannot exist *ad infinitum* because of the impossibility of an actual infinite. Therefore, he argued, the world requires an initial 'generator' (*muhdith*) who could generate it *ex nihilo*.

The famous Muslim physician and Neoplatonist al-Râzî,[24] on the other hand, seems to have adopted Plato's notion (in the *Timaeus*) that time is 'the moving form of eternity', as well as Plotinus' notion that time is eternal; therefore he refused Aristotle's view of the unreality of time.

The influential philosopher al-Fârâbî,[25] on his part, focused on the metaphysical aspects of time. He also adopted Aristotle's view when he said: 'only the circular motion is continuous and time is related to this motion' ('Abdul-Muta'âl 2003: 113). But similarly to Ibn 'Arabî, al-Fârâbî apparently also believed that the world is contingent or 'possible to exist' before its actual existence. Otherwise, if it were 'impossible' it would not exist, or if it were 'necessary', it would have always been. Then he stresses that the world *as a whole* is in continuous formation (*takwîn*) and corruption (*fasâd*) 'in no time', while the parts of the world are forming and deforming in time ('Abdul-Muta'âl 2003: 115). This outlook is also similar to Ibn 'Arabî's view (see section 2.3). To explain this important point we give the following illustration. If a young physicist was asked to describe the general state of a mountain, he or she might end up with some equations without any reference to time, because the mountain is rigid. But if we ask him or her to include in his study the fact that the mountain is part of the Earth which is rotating around its axis and orbiting around the Sun, and also the fact that the atoms in the rocks never rest, or even the motion of the insects and other animals that might be living there, as well as the motions of the wind, waters ... etc. – then in that case the physicist might need to invent some new mathematics in order to be able to include the time parameter properly in his or her equations, even after making many approximations. So because we live 'inside' the world we feel time, but the entire world itself is out of time.

Many other schools of Islamic thought have speculated on the issue of time. It is good to notice, however, that in the Qur'an itself Allah never makes any direct reference to the usual Arabic word for 'time' (*zamân* or *zaman*), although many other time-related terms later explored by Ibn 'Arabî – such as the year (*sana* and *'âm*), month (*shahr*), day (*yawm/nahâr*) and night (*layl*) – were mentioned very often in the Qur'an, in addition to some divine Names that are related to time such as the First (*al-awwal*), the Last (*al-âkhir*) and the Age (*al-dahr*).

As already noted in our Introduction, with all the hundreds of books and recent studies that have been written about time in Islamic philosophy, both in Arabic and in other languages, it is very strange almost none has ever focused on Ibn 'Arabî. Many scholars have studied and compared the different theological, philosophical and physical views of time and existence, so briefly summarized above. However, none of these studies has ever, to the best of our knowledge, treated Ibn 'Arabî's unique concept of time, although – as we shall discover – it is actually the key to understanding his controversial theory of the oneness of being. The reason for this strange neglect could be not only his difficult and symbolic Sufi language but also the fact that his concepts are intentionally dispersed throughout his many writings and not plainly stated in one place, as most other authors do. In fact, Ibn 'Arabî mentions in the *Futûhât* [I.141.13] that he wrote a treatise with the title of '*al-zamân*' ('time') where we would expect to read at least an extensive summary of his view of time. However, apparently this precious work has been lost, although *al-Futûhât al-Makkiyya* seems to include most of his doctrine regarding time and other related cosmological issues.

1.8 Time in Western philosophy

Aristotle's notion of circular time, based on an eternal (uncreated) universe, could not generally be accepted by most theologians of the three Abrahamic religions – Islam, Christianity and Judaism – in so far as they considered time to be linear, with a definite created beginning and end. St Augustine, and later Thomas Aquinas, objected to Aristotle's belief that time is circular, insisting instead that human experience is a one-way journey from Genesis to Judgement, regardless of any recurring patterns or cycles in nature. This latter view was later adopted by Newton in 1687, when he represented time mathematically by using a line rather than a circle.

As we noted above, there was already an earlier debate in Greek philosophy as to whether time exists objectively, or is just constructed by our minds. Puzzled about time, St Augustine concludes that time is nothing in reality, but it exists only in the mind's apprehension of that reality (*EP*, 'Time', VIII: 126). On the other hand, Henry of Ghent and Giles of Rome both said that time exists in reality as a mind-independent continuum, but is distinguished into earlier and later parts only by the mind. Isaac Newton considered time (and space) as an independent quantity that exists and flows regardless of matter or mind, a view which Leibniz strongly criticized. Leibniz argued that if space is distinct from everything in it, it would have to be completely uniform and homogeneous; thus he reached the conclusion that it is unreal and relative, in anticipation of Einstein's Relativity, though he never put that insight into the form of mathematical equations (Ross 1984: 47).

Newton also rejected Aristotle's linkage between time and motion, when he said that time is something which exists independently of motion and which existed even before God's creation. He argued that time, and space are an infinitely large 'container' for all events, and that the container exists with or without the events: this is called the 'absolute' theory of time. Leibniz, who adopted the relational view, objected to that and argued that time is not an entity existing independently of events.

In the eighteenth century, Kant said that our mind structures our perceptions so that space always has a Euclidean geometry, and that time has the structure of the infinite mathematical line (Kant 1998: 158–176). This view, however, lost its mathematical support with the discovery of non-Euclidean geometries in the 1820s.[26] In his *Critique of Pure Reason*, Kant presented, in the first antinomy, two equally plausible arguments that at the end lead to opposing conclusions: the first shows that the world had a beginning in time, while the other shows that the world had no beginning in time. For the second proposition, if we suppose that the world had no beginning in time, then this means that, at any particular moment of time, an infinite number of events have passed – but infinity may never be completed. On the other hand, if the world had a beginning in time, then all previous times before that beginning have been blank, and there is no any specific reason why the world should have begun at this time in particular. (The same argument was earlier used by Leibniz in support of his relational theory of time.) We shall see in section 2.3 how Ibn Arabî gets out of these riddles by criticizing the questions themselves and asserting instead that Allah created (and is continuously and recurrently creating, see section 5.6) the world and time together (Kant 1998: 462–463, 470–476, 490–495, 525–528, 536–538).

In a famous article in *Mind*, McTaggart also argued about the unreality of time. Events, for McTaggart, are capable of being ordered in two ways: either as past-present-future, which he calls A-series, or as 'earlier than'-'later than', which he calls B-series. He then argues that A-series is contradictory, and that B-series does not give all that is essential to time, because time also involves change. The A-series view of time is contradictory because, as McTaggart argues, it will lead to the fact that each event can be described (at different times) by future, present and past, because events are always flowing from the future to the past through the present, so at some times an event may be future then it becomes present and then past; while in the B-series it is always before other events or after other events, no matter whether those events are future, present or past (Dyke 2002: 137–152).

On the other hand, there has also been a great debate as to whether time is continuous or a discrete quantity. Most Western philosophers think of time as a continuous quantity, but after the advent of Quantum Mechanics the idea of quantum time was revived, although Quantum Theory itself does not consider time to be 'quantized' (Mehlberg 1971: 16–71).

1.9 Time in modern science

Quantum Field Theory and General Relativity are the most well-established modern fundamental theories of physics. According to these theories, space-time is a collection of points called 'space-time locations' where physical events occur. Space-time is a four-dimensional continuum, with physical time being a distinguished, one-dimensional sub-space of this continuum, but no longer a separate entity nor space: space and time are always taken together as one entity.

In 1908, the mathematician Hermann Minkowski, Einstein's teacher, was the first person to realize that space-time is more fundamental than time or than space alone. As he put it:

> The views of space and time which I wish to lay before you have sprung from the soil of experimental physics, and therein lies their strength. They are radical. Henceforth space by itself and time by itself are doomed to fade away into mere shadows, and only a kind of union of the two will preserve an independent reality.
>
> (Pais 1982: 152)

The metaphysical assumption behind Minkowski's remark is that what is independently real is what does not vary from one reference frame to another. It follows that the division of events into the past ones, the present ones and the future ones is also not independently real.

In contrary to the classical Newtonian view, time intervals depend greatly on the observer's frame of reference. In classical mechanics, and on the basis of common sense, if the time interval between two lightning flashes is 100 seconds on someone's clock, then the interval also is 100 seconds on your clock, even if you are flying by at an incredible speed. Einstein rejected this piece of common sense in his 1905 special theory of Relativity when he declared that the time interval (and the distance) between two events depends on the observer's reference frame. He says that every reference-body has its own particular time; unless we are told the reference-body to which the statement of time refers, there is no meaning in a statement of the time of an event (Einstein 1920: ch. 9). Thus each reference frame (or reference-body) divides space-time differently into its time part and its space part.

1.9.1 Curved time and the Big Bang

In 1922, the Russian physicist Alexander Friedmann predicted from General Relativity that the universe should be expanding. In 1929, Hubble's measurements of the redshift confirmed this prediction. Eventually astronomers concluded that about 12 to 15 billion years ago the universe was in a state of infinite density and zero size; this is the Big Bang theory referred to earlier in this chapter. So we might ask the simple question 'What was there before the Big Bang?' Astronomers, however, showed that the entire universe, including time and space, was created in the Big Bang, and because of the extremely high density of matter at that instant, the gravitational force was immense and the space-time was curved, or encapsulated around the point from which the Big Bang was ignited.

In physics and modern philosophy, descriptions of the Big Bang often assume that a first event is also a first instant of time and that space-time did not exist outside the Big Bang. But it is not clear if it is correct to call the Big Bang an 'event', because events must have space and time co-ordinates, but space-time started with the Big Bang itself. However, for the first time in science there is a mathematical description of the ontological relation between time and the universe. We shall see that this description is in good agreement with Ibn 'Arabî's own views on time (section 2.3).

However, there are serious difficulties in defending the Big Bang's implications about the universe's beginning. Current theories do not have any claims as to what might have happened before the Planck era (10^{-43} seconds after the Big Bang). It is expected that the theory of Quantum Gravity might provide information about that, and it may even allow physicists to speculate on what caused the Big Bang. But until now this area remains solely in the domain of theological and metaphysical speculation.

1.9.2 The arrow of time

Unlike physical space, physical time is inherently directional: it flows in one direction, from the future to the past; this is a necessary truth. In thermodynamics the arrow of time for the world is expressed through what is called the 'entropy', which is a description of the degree of order of a system; a highly ordered system is said to have smaller entropy, and vice versa. The world's entropy is always increasing (i.e. its order is decreasing): this is a free 'one-way ticket', and one has to pay dearly for the return. For example, the process of mixing hot water into cold water to get warm water is never reversed, though in principle we may think of some complicated machine that can do the reverse. The arrow of an irreversible physical process is the way it normally goes, the way it normally unfolds through time.

The problem with the arrow of time is that the variable time is symmetric in most equations of physical laws. This means that, if the variable t is replaced by its negative $-t$ in those laws, the result is still a law; the basic equations are unchanged. Some scientists theorize that the cosmological arrow of time will one day reverse direction when the force of gravity will halt the further expansion of the universe and start a collapse to its initial state, just like a movie played backwards (Price 2002: 19–56).

1.9.3 Time travel

One of the most fascinating consequences of the theory of Relativity is that it allows travel through time, just as one travels through distance. This has been employed by science fiction to produce many interesting films and science fiction stories.

Although reversing the direction of the 'arrow of time' still seems to be experimentally impossible, this happens quite often in dreams and in remembering past events. However, philosophers have been more interested in travel in physical time than in psychological time. On the other hand, we can also 'really' look at the past any time: by looking at the stars, where we actually see how they have been thousands and millions of years ago when the light that we see now was actually emitted by them. But this is still not like travel in physical time, which is conceivable now only in theoretical cosmology.

Although travel in time is possible and allowed according to the equations of Relativity, in many cases it violates logic and causes obvious paradoxes. There are, however, different types of time travel, some of which are trivial. If you get on a plane on the Earth's surface and travel west, you will cross a time zone and instantly go back an hour, but all you have done is to change your reference frame. Also, if your body were quick-frozen in the year 2000 and thawed in 2050, then you would travel forward 50 years in clock time but only a few seconds of your biological time. This is, however, just a case of biological time travel, not a case of physical time travel.

One possible, and more genuine, way to travel through time is to fly at enormous speeds close to the speed of light. Einstein showed that travel forward in physical time is possible relative to the time of those who move more slowly than you. With this kind of relativistic time travel, you cannot return to the old present, but you can conceivably be present at the birth of your great-grandchildren. Travel backward in physical time is also possible only if nothing that has happened gets changed. For example, you cannot go back in time and prevent your parents from having any children (Arntzenius and Maudlin 2002: 169–200).

Another kind of time travel is caused by the curvature of time due to extreme gravity. If you fell into a black hole, then you would travel to a time after the end of the universe, as measured in a reference frame tied to Earth. Unlike the time travel in science fiction movies, this kind of relativistic time travel to the future is continuous, not abrupt. That is, as you travel to the future, you exist at all intervening times according to the stationary Earth clock. You do not suddenly 'poof' into existence in the year 4500; you existed during their year 4499, but your spaceship had not yet landed.

Going back to the past is probably possible, but there are significant difficulties yet to be overcome before we can be sure. In recent decades, mathematicians and theoretical physicists have described time machines, or at least universes containing backward time travel, that are consistent with Einstein's equations of general relativity. However, Stephen Hawking believes that all these time machines are ruled out by the laws of General Relativity.

For Ibn 'Arabî, time travel is possible and easily attainable without any paradoxical consequences. Such time travel, however, has no physical or biological effects on the traveller – see the discussion in section 2.7 below.

1.9.4 Quantum time

Regarding the question of 'instants' of time, time being a linear continuum implies that there is a non-denumerable infinity of them. This means that between any two instants there is a third; time is continuous. However, for times shorter than about 10^{-43} seconds, the so-called Planck time, science has no experimental support that time holds its continuousness. But physicists agree that General Relativity must fail for durations shorter than the Planck time,[27] though they do not know exactly how and what is the substitute.

The idea that space or time (space-time) could be discrete has been recurring in scientific literature recently, but its origins go back to ancient philosophies.

The new concepts brought about by Quantum Mechanics (e.g. the concept of indeterminacy or the uncertainty principle) suggested that space-time could be also quantized like energy. This was reinforced by the discovery of ultraviolet divergences in Quantum Field Theory (Zee 2003: 145–151), though many of the strange quantum concepts soon became acceptable aspects of continuum physics. In the 1980s, powerful computers inspired some new discrete thinking in physics. Complicated mathematical simulations performed on these super-computers paved the way for lattice theories to be applied to Quantum Mechanics, and included Quantum Gravity. In Quantum Gravity, Planck's length is a minimum size beyond which no accurate measurements can be performed.

Hawking, however, sees no reason to abandon the continuum theories that have been so successful. But it may be possible to invent a discrete structure of space-time without abandoning the continuum theories if the discrete-continuum duality can be resolved, just as the wave-particle duality has been resolved by Quantum Mechanics.[28]

The practical methods of the quantization of time in modern scientific theories are based on some complicated mathematics such as lattice theories and cellular automata that are beyond the scope of this introduction (Wolfram 2002: 771). But it is good to note here that Ibn 'Arabî's quantization of time, as we shall see in section 2.8, is unique and is based on a broader cosmological view (of the oneness of being) such that discreteness and continuousness are special cases of it (see also section 7.5).

1.10 Introducing Ibn 'Arabî's view of time

As we shall see in Chapter 2, Ibn 'Arabî considers time to be imaginary and without real existence; it is only a tool used by the mind to chronologically arrange events and the motion of the heavenly spheres and physical objects. Ibn 'Arabî then distinguishes between two kinds of time: 'natural time' and 'para-natural time'. He also explains that the origin of this ultimately imaginary time is from the two forces of the soul: the active force and the intellective force.

Despite time being imaginary, Ibn 'Arabî considers it as one of the four main constituents of nature: *time*, *space*, the *monad* (*al-jawhar*), and the *form* (*al-'arad*). Like some modern theories, Ibn 'Arabî also considers time to be cyclic, relative and inhomogeneous.

Ibn 'Arabî then gives a precise definition of the 'day', the 'daytime' and the 'night' and generalizes that, in relation to all (real and imaginary) orbs or spheres, every orb has its own 'day' and those days are measured by our normal day that we count on the Earth.

On the other hand, Ibn 'Arabî gives special importance to the cosmic 'Week', and says that the seven cosmic week-Days are unique and not alike. Saturday (*al-sabt*) in particular has a special importance, because he considers it to be the '*Day of eternity*', so that the observable week days, including Saturday itself, are therefore happening in Saturday! This initially may look rather confusing, but it should become easier to understand, especially after we explain Ibn 'Arabî's view of the re-creation principle and his theory of the oneness of being which we discuss in detail in Chapter 5.

Finally, what is very important and unique about his view of time is that Ibn 'Arabî considers time to be discrete: there is a minimum indivisible 'day' or 'time' – and thus, surprisingly, this 'day' is equal to the normal day itself which we live and divide into hours, minutes, seconds and much less than that! This conception at first looks very strange and ambiguous, but, in order to explain this, Ibn 'Arabî introduces three kinds of days, depending on the actual flow of time that is not so uniform and smooth as we ordinarily imagine. The key point here is that Ibn 'Arabî stresses that, according to the Qur'an, *only one 'event'* should be happening every 'day' (of the actual days), and not many different events as we observe. To achieve his deeper understanding of this key Qur'anic expression, he reconstructs the underlying reality of the normal days in a special way from the different days of the actual flow of time, as we shall discuss further in Chapter 4.

Also on the basis of a number of key verses in the Qur'an, Ibn 'Arabî says that the world ceases to exist instantly and intrinsically the next moment right after its creation, and then it is re-created again and again. We shall see that Ibn 'Arabî's view of time and how it flows is precise and unique; it has never been suggested or discussed by any other philosopher or scientist. This distinctive cosmic vision of 'ever-renewed creation', when added to the understanding of the actual flow of time based on the three kinds of days alluded to above, can be used to build a new unique model of the cosmos which we shall discuss in Chapter 6 and we shall discuss some of the consequences of this model in Chapter 7.

2 General aspects of Ibn 'Arabî's concept of time and days

Time, if you investigate what it comes down to, is something verifiable;
Yet it is known (only) through (human) imaginations (*awhâm*),
Its power is like Nature in its effects,
Though the essence of both is non-existing.[1]
Through it all things are determined,
While it itself has no existing essence by which it could be judged.
[I.291.1–7]

We have seen in the previous chapter that time is one of the most important perennial problems in physics and cosmological philosophy. For the same reason, we find that Ibn 'Arabî considers that a good understanding of time is the prime key to human spiritual realization. In this chapter we shall look at the various relevant aspects of Ibn 'Arabî's view of time. We want to give here some general descriptions of his ideas on time, and we shall focus on the important issues in the following chapters. For this reason there are many cross-references to following chapters where these ideas are explained in more detail.

2.1 What is time?

To start with, Ibn 'Arabî declares that time is an imagined attribute that does not exist on its own; it has no separate physical or non-physical entity. He argues that 'time in relation to us is like eternity in relation to Allah, and since eternity is a negative attribute[2] that does not exist on its own, so time in relation to the contingent world or the entire cosmos is (also) an imagined attribute that does not exist' [I.291.28].

In his major book *Al-Futûhât Al-Makkiyya*, he says: 'we showed in this book and in our book *The Time* (*al-zamân*) (OY, no. 838) that time is something that has no (real) existence' [I.490.17]. Although this last book is not found today, Ibn 'Arabî's concept of time is developed in quite detailed fashion in the *Futûhât*, though it is scattered all around the book and not placed in specific parts, including even those chapters 59, 291 and 390 whose titles relate directly to time.

The concept of time is needed to compare the sequence of events or motion, but real existence is attributed only to the thing that actually moves, not to the abstractions of motion, time or space in which motion is observed:

> And time and space are also a consequence of natural bodies, but time is something imagined that does not exist (in itself), but is introduced by the motion of orbs and localized things when we ask about them 'when'. So time and space do not exist in reality, but existence is to the things that move and still.
>
> [II.458.1]

This is not only to say that 'motion', 'space' and 'time' do not have real physical existence, but they do not even exist separately in an abstract way: their existence is a mere illusion; it is only a projection of the human imagination (*wahm*).

It is not very easy to deny the existence of time, space and motion, since they are widely encountered in our experience of everyday life. However, Ibn 'Arabî is not the first one to propose this. We shall see below that Aristotle gives a very simple proof that time is not real. The real existence of motion and space, however, are far more unusual and intricate to disprove. Perhaps only Zeno (b. *c.* 488 BC) was brave enough to postulate the illusion of motion, and he composed many related riddles that are still logically unsolvable. The main idea behind Ibn 'Arabî's mysterious conceptions here is his controversial theory of the oneness of being, which we shall discuss throughout this book, especially in Chapter 5. But, clearly, if we suppose that the 'real' existence in the world is uniquely one, there would be no meaning to motion, and hence to time and space; or at least they would have to be redefined. We shall discuss in Chapter 7 that Zeno's paradoxes can be easily resolved according to Ibn 'Arabî's profound view of time under his theory of the oneness of being.

Coming back to time, we can say that it can be very easily shown that it is in fact imaginary. Aristotle says in his *Physics* that 'time consists of two parts, one of which has existed (and gone, i.e. the past), the other does not yet exist (i.e. the future), so how can something exists which is composed of what does not exist?' (Lettinck 1994: 348)

So if there is a *real* existence to time it will be in the *present* (the 'now'), not the past or the future. Aristotle then gives another argument that also the 'now' is not time. The present is not time, but it is rather a point in imaginary time, like a point on the line; although the line is composed of points, still each point is not a line.[3] Likewise time is the sum of all present moments that exist only one by one, and each one present moment (alone) is not time. Time therefore is the mind's projection on the continuous presence from the future to the past through the present.

Similarly, Ibn 'Arabî gives a straightforwardly profound meaning of time. In the title of chapter 390 of the *Futûhât*, he says: 'the time of a thing is its presence' [III.546.16]. Then he explains that the time of the Lord is the 'servant' and the time of the servant is the Lord (*al-rabb*) [III.547.31], because the Lord deserves to be given this name by the servant, since He would not be called 'Lord' if there are no servants to worship Him; likewise the servant deserves his name by (his relation to) the Lord. In the same way, when we say for example: ''Amr is the son of Zaid', this means, according to Ibn 'Arabî, that the time of the fatherhood of 'Amr is the sonhood of Zaid, and vice versa. Or in Ibn 'Arabî's own words: 'the time of the father is the son, and the time of the son is the father' [III.547.36]. That is why Ibn 'Arabî gave this chapter (390) the title 'The time of a thing is its presence, but I am out of time and You are out of time, so I am Your time and You are my time'. This means: 'I am Your presence and You are my presence.'

'Time' in the usual common sense is actually a tool used by our perception to classify the events or motion of objects chronologically; it would not have any meaning without motion or change. This is why we do not feel time while we are asleep; we have to look for some kind of a standard reference motion (the Sun, the Moon, the stars or a watch) in order to realize how much of time has elapsed since we went into deep sleep. Time, therefore, has no real absolute meaning; it is only used relative to something in order to describe its state of existence. This is why Ibn 'Arabî sometimes uses the words 'time' and 'state' as synonyms, as when he says: 'as you like (you can) say from the time of its existence, or the state (*hâl*) of its existence' [II.281.11].

So the real meaning of time is reduced to the existence of the world in the 'present moment', which has no duration or extension, because the future and the past are mere imagination. If we know that, Ibn 'Arabî declares, there is no problem to go along with people and say that 'time' is the daytime and night, or that it is a duration taken by the motion of objects, or it is comparing an event to another when someone asks about it by 'when?', because these definitions have been widely used and they are correct in relation to time, in the common sense [III.548.7]. As Ibn 'Arabî says in chapter 59 of the *Futûhât*, which is titled 'On knowing existing and assumed time', people have used the word 'time' (*zamân*)[4] in many different ways: most philosophers, for example, use it as the duration taken by the motion of orbs. Muslim theological scholars, on the other hand, use it to order events sequentially. But generally *zamân* in Arabic means the night-time (*layl*), and the daytime (*nahâr*). Also Ibn 'Arabî himself sometimes uses the word *zamân* to mean the daytime and night [I.141.5]. But this is more than just a convention, because, as we shall see below and in the following chapter, the (divine) 'day' (*yawm*) is for him the main indivisible unit of time. So if time has any real existence it exists as the divine 'days' (of each real instant), not as the hours or seconds we conventionally use. It is also worth mentioning here that, unlike the day (*yawm*) or daytime (*nahâr*) and the night (*layl*), the word 'time' (*zamân* or *zaman*) was never used in the Qur'an.

Despite the fact that he considers time to be imagined and having no real existence, Ibn 'Arabî stresses that it is one of the four 'mothers (fundamental principles) of existence': the formable monad (*al-jawhar al-suwarî*),[5] the accidental form (*al-'arad*),[6] time (*al-zamân*) and space (*al-makân*).[7] Everything else in the manifest world is combined of these four parameters [III.404.22]. He also argues that those four parameters – together with another six categories that are derived from them: *fâ'il, munfa'il, idâfa, wad', 'adad, kayf* – are enough to describe the state of everything in the world. Together these make up the familiar ten Aristotelian categories: i.e. substance (*jawhar*), quantity (*kamm*), quality (*kayf*), relation (*idâfa*), time (*matâ*), place (*ayna*), situation or position (*wad'*), possession (*lahu*), or state (*jidda*), passion (*yanfa'il*) and action (*yaf'al*) – although the meaning of *jawhar* here is of course radically different from its usual Aristotelian usage, reflecting in this case the kalam inspiration of Ibn 'Arabî's terminology (*EI²*, VI: 203, 'Al-Makûlât', and: *EP*, II: 46, 'Categories'). Yet those four 'mothers of existence', including the formable monad, in Ibn 'Arabî's distinctive conception of the oneness of being, are nothing but imaginary forms or reflections of the unique 'Single Monad' (*al-jawhar al-fard*) which is the only thing that can be described as having a real existence: all other things in the world are different forms of this Single Monad, including 'vision and the visible, hearing and the heard, imagination and the imaginable, thinking and the thinkable, . . . etc.' [III.404.12]. This latter concept reflects Ibn 'Arabî's controversial theory of the oneness of being.

The importance of understanding the reality of time is, therefore, to provide the link between the *actual* unity of this Single Monad and the *apparent* multiplicity of the witnessed world. That is why Ibn 'Arabî says that:

> the knowing of time is a noble knowing through which eternity (*al-azal*) is truly known. . . . But only the Solitary Sages[8] among 'the True Men' know it (eternity). This (reality) is known as 'the First Age' (*al-dahr al-awwal*) or 'the Age of ages' (*dahr al-duhûr*). From this eternity (*al-azal*),time comes into existence.
>
> [I.156.34–157.1]

Therefore, we may summarize that time as we ordinarily experience it is defined by motion, and motion is defined by the different positions of the formable monads, and those monads are different states (times or instances) or forms of the Single Monad, which alone has a real existence.

2.2 Physical time and spiritual time

Ibn 'Arabî distinguishes between two kinds of time: physical or 'natural time' (*zaman tabî'î*) and spiritual or 'para-natural time' (*zaman fawq-tabî'î*). The first is used to compare the motion of bodies and orbs, while the latter is used to compare the changes in spiritual states, such as realizing and knowing. He explains that the existence of time does not necessarily require the existence of matter [IV.337.5], because there is time that is associated with material motion that is under the effect of Nature, and time that is associated with immaterial motion that is above the effect of Nature: i.e. in the spiritual world. Thus he says that 'you should know that some of time is above Nature and some of it is below Nature' [I.377.12], and he explains further by saying that the time that is under

Nature 'is defined by the motion of orbs . . . and the time that is beyond Nature is defined by (spiritual) states' [I.477.12]. So when Ibn 'Arabî says: 'and the origin of the existence of time is Nature, whose state is below the Universal Soul and above the Universal Dust' [III.548.19], this actually refers to natural time which is used to compare the motion of bodies and orbs – while spiritual time is used to compare the change of spiritual states, and its origin preceded the existence of the physical world.

We can fairly say that Ibn 'Arabî's conception of natural time is the time known in physics and cosmology, and that his spiritual time resembles what is called by many modern philosophers 'psychological time'. The psychological time is our feeling of time's passage even when everything around us is standing still, including ourselves: we feel this time because our inner state is continuously being updated through changes in our consciousness, unless we fall into deep sleep, as happened for example with the 'people of the cave' (*ahl al-kahf*) who stayed over three hundred years in deep sleep, yet when they awoke they thought it was like a day or part of a day (Qur'an 18:9–25, see also [II.9.21]).

Physical time seems to flow uniformly and continuously (at least locally), while psychological time depends greatly on the mood, as Ibn al-Fârid suggested nicely in one of his poems:

> With her for me a year is like a glimpse,
> and an hour of parting for me is like a year.
> (Mahmûd 1995: 344)

Also Ibn 'Arabî says that 'minutes are years while sleeping' [IV.337.1], and we shall discuss this relative aspect of 'para-natural', psychic time in section 2.7.

2.3 The origin of the world

Many people ask questions like 'what is the age of the universe?' or 'when did the world begin?' And many cosmologists go along with these questions and give estimates for the age of the physical universe (today, usually about 15 billion years). Any answers to such questions will quickly lead to a modern version of the still-ongoing debate between Plato's and Aristotle's schools already mentioned in Chapter 1: i.e. whether time was created in or before the world, or vice versa; or whether they are both eternal. Many riddles and paradoxes quickly emerge out of this debate. For example, one may ask: if the world started at a certain point of time, why God chose that time in particular? Could the world have been created ten minutes before or after that designated time? And what was God doing or what was happening before the beginning of the world?

Ibn 'Arabî, however, shows that such questions are meaningless, because the world is created out of time, and time itself is part of that created world. Allah did not create the world in time, because nothing existed 'before' the world apart from Him Who is also out of time, and therefore the creation of the world can not be compared to other events in time. The existence of the Creator precedes the existence of the world *logically*, and not chronologically, so it is like when we say 'the day starts when the Sun rises': there is no duration of time that separates sunrise from the start of the day [I.100.26, *Al-Masâ'il*, no. 25], but logically the day would not start if the Sun did not rise. The existence of Allah, Who is Pre-Existent (*Qadîm* or *Azalî*) and Self-Existent (*Qayyûm*), is a precondition – not a cause – for the existence of the world. Therefore, because the world was created out of time (not 'in' time), the above questions are invalid. All such questions use time phrases that have no meaning outside of that time which appeared in or with the world and not before it.

Ibn 'Arabî showed this clearly by explaining that:

> The fact of the matter is that the existence of the Real is not determined (temporally or causally) by the existence of the world: not temporally before, with, or after, because temporal or spatial precedence with relation to Allah is confronted by the realities confronting whoever speaks about it factually – unless they say something by the way of illumination, as had been said by the Messenger, peace be upon him, or it was expressed in the (divine) Book. For not everyone is able to experience the unveiling of these realities. We can only say that the Exalted Real exists by Himself and for Himself; His existence is absolute, is not confined by any other than Him, and is not caused by something nor is He the cause of anything – But He is the Creator of causes and results, *the King, the Most-Holy One* (59:23), Who always is and has been.
>
> The world exists *through Allah*, not by itself or for itself. Therefore the existence of the Real Who exists by Himself is a determining condition for the existence of the world, which would not exist at all without the existence of the Real. And since time can not exist without the existence of the Real and the (divine) Source (*mabda'*) of the world, therefore the world comes to exist 'in other-than-time'. So actually we can not say, in the true reality of things, that Allah existed before the world – because it has been established that 'before' is a time phrase, and there was no 'time' (before the existence of the world). Nor can we say that the world existed after the existence of the Real, since there is nothing (other than the Real) 'after' or 'with' the existence of the Real, because He caused (everything else) to exist and is making it and originating it (in the words of a famous hadith) 'while there was no thing (with Him)'.
>
> So as we said, the Real exists by Himself and the world exists *through* the Real. But if someone governed by his imagination (*wahm*) should ask 'When was the world (created) after the Real?' we would say that 'when' is a time-question. But time belongs to the realm of relations (*nisab*, see: *SDK*: 35) and (as such) is created by Allah, but not like the creation of existence, because the realm of relations is created by (our human) 'estimation' (or consideration: *taqdîr*), not by the creation of what exists . . . Therefore this question is not valid. So you should be careful how you ask, and do not be

veiled by the tools of (conventional human) expression from actually realizing and fully comprehending these realities in yourself.

So the only thing left is: (1) a pure and absolute Existence – not (one coming into existence) after non-existence – and that is the existence of the Real, may He be exalted! And (2) an existence (that only comes to be) after the non-existence of the essence of that existent thing itself – and that is the existence of the world. So there is not any comparability or (co-extensive) extension between those two existences, apart from that imagined, presumed one that is removed by (true) knowledge. So nothing is left of that (falsely supposed comparability of God and the world) but absolute Existence (of God) and determined existence (of the world), Active Existence (of God) and passive existence (of the world). This is what is given by the realities, and peace (i.e. that's all!).

[I.90.9, see also *Al-Masâ'il*, no. 90]

'The world' for Ibn 'Arabî is both spiritual and material, and as we have seen above there are two corresponding kinds of time, spiritual and physical. The spiritual world preceded the creation of the material universe (nature) as we know it (stars and planets), so there was spiritual time before the creation of the physical world. He indicates that spiritual time is necessary to describe the relation between spirits and the divine Names before the creating of the physical world. So in this respect Ibn 'Arabî does respond to the above-mentioned questioning about what was 'before' the creation of the physical world,[9] but he still considers such questioning invalid beyond these two created realms of the world. In fact Ibn 'Arabî explains that the world has three distinctive ontological 'levels': *'alam al-mulk* or *'alam al-shahâda*, which is the visible world; *'alam al-malakût*, which is the realm of meanings; and *'alam al-jabarût*, which is the all-encompassing realm of the divine 'Imagination' (*barzakh*) [I.54.15, II.129.16]. But here and elsewhere he sometimes adapts the simplified reference (drawn from the Qur'an) to the two domains as the 'visible' and 'invisible' (see *SPK*: 14, 93, 129, 218, 223, 342, 360–361, 376, and also *Al-Masâ'il*, no. 150).

Actually, Ibn 'Arabî considers the age of the world (as spiritual and material together) to be infinite from both directions: i.e. it has always been and it will always be; it is eternal without beginning and eternal without end. However, this is not saying that the material (and even the spiritual) world is eternal, but the world has some sort of pre-existence in the foreknowledge of Allah, and Allah's knowledge is eternal in both directions. In addition, Ibn 'Arabî also considers those two ends to coincide with each other, so time as a whole is like a circle that can not be described to have a beginning or an end, but when we set a point (the present) and a direction (to the past or future) on this circle we define a beginning and an end [I.387.32, III.546.30]. We shall discuss the concept of circular and cyclical time in section 2.10.

Ibn 'Arabî – following Ibn Sînâ's familiar theological categories (Nasr 1964: 173–274) – divides all things, in terms of their type of existence, into three inclusive categories: necessary, possible (or 'contingent': *mumkin*), and impossible

[II.293.1, II.575–576]. Only Allah's existence is described as 'Necessary' or Self-Existent, while absolute non-existence is impossible. The world, on the other hand, is called 'possible' because it is *possible* to exist, but in order to actually come into existence it needs a determining cause (*murajjih*) who has to be pre-existent and self-existent, or none other than Allah.[10] Therefore, the world (the possible) is originally non-existence (but not absolute non-existence, only a non-existence that is possible to exist) and it is always in need of Allah in order to bring it into existence. Thus it can be said – as in the long quotation above – that it exists *by* or *through* Allah, and not through itself. The difference between absolute non-existence and possible non-existence is that the latter exists (even before it comes into real existence) in Allah's Knowledge, and this foreknowledge is eternal because His knowledge is not other than Himself [I.300.32, II.114.15] (see also section 2.7).

So the world, spatial and temporal, eternally existed or was determined in Allah's knowledge, but it is continuously brought into real existence *ad infinitum*:

> So when Allah brought the entities (into existence), He brought them for them not for Him. But (for Him) they are still as they were on their spatial and temporal states, with their different time and space (co-ordinates). So He reveals to them their entities and states little by little infinitely and successively. So the issue for Allah is one, as He said: *and Our Command is but one, as the twinkling of an eye* (54:50), and multiplicity is (only) for the countables themselves.
>
> [I.162.22]

However, despite this pre-existence, we can not say that the world is eternal and only developed from one state to another. It is not exactly clear how Henry Corbin concluded that 'there is no place in Ibn 'Arabî's thinking for creation ex nihilo, an absolute beginning preceded by nothing' (Corbin 1969: 200), when Ibn 'Arabî started the *Futûhât* by saying: 'Praise be to Allah Who created things after (its being) non-existence' [I.2.3]. In *Al-Durrat Al-Baydâ'* he also declares that 'the dependent existence (the possible) may only be after non-existence, otherwise it would not be a "possible" whereas it is possible, and it would not be an existence by the Self-Existent, while it is in fact an existence by the Self-Existent Who caused it to exist' (*Al-Durrat Al-Baydâ'*: 133).

If Corbin means what we explained in the previous paragraph – i.e. that everything existed in the foreknowledge of Allah even before it really existed in the world – then we have to stress the difference between the essence or entity (*'ayn*) of a thing in God's Knowledge and its actual existence. William Chittick devoted a full chapter in his book *The Sufi Path of Knowledge* to explaining this important concept in Ibn 'Arabî's ontology (*SPK*: 77–143). The entities of the world are in Allah's foreknowledge eternally, but they are brought into existence – after they were not existing – one after the other. This is a very important distinction. Ibn 'Arabî continues by explaining that:

> Everything is in need of Allah, the Exalted, for the *existence* of its essence (or entity: *'ayn*), not for its essence, because its immutable essence (*'aynuhu al-thâbita*, most widely translated as 'permanent archetype', see *SPK*: 83) is not determined in its immutability; it is not determined by a determiner, for there is no determination in eternity . . . so the existence of the possible may only be after non-existence, which means it was not, and then it is.
>
> (*Al-Durrat Al-Baydâ'*: 133, see also *Al-Masâ'il*, no. 143)

Actually we shall see later in section 5.6 that this intrinsic need by the 'possible' – i.e. of all the realms of creation – for Allah to bring it into existence continues to be necessary at every single moment, because the world is continuously brought into existence in ever-renewed forms (i.e. it is constantly '*re-created*').

However, it is still not easy for the human mind to imagine the existence of the created world (*al-muhdath*) and the eternal existence of Allah, the Eternal or Pre-existent (*al-qadîm*) without a reflection of some time separation. And even for Ibn 'Arabî the issue is not yet closed. As he suggested in the long passage quoted above, somehow understanding this mysterious point seems to be beyond normal human perception and requires a divinely inspired knowledge accessible only to rare individuals with the very highest spiritual attainments. This same difficulty caused many Muslim philosophers and other theologians to continue to speculate on many theories that Ibn 'Arabî disagrees with (*Kitâb Al-Azal*: 8).

Ibn 'Arabî explains further the relation between the existence of God and the existence of the world in chapter 59 of the *Futûhât*, which is the same chapter in which he talks specifically about the topic of time. Ibn 'Arabî's argument presented at the beginning of this chapter is extremely complicated and very difficult to understand, even in its original language. However, because of its importance, we are obliged to translate it here, with some necessary explanations in parentheses.

At the beginning of chapter 59, after the opening poem quoted at the very beginning of this chapter, Ibn 'Arabî says:

> You should know first that Allah, the Exalted, is *the First* (*al-awwal*), and there is no firstness (*awwaliyya*) to anything before Him nor to anything else – whether that exists through Him or independent of Him – with Him: but He is *the One* (i.e. the Unique; *al-wâhid*), Glory be to Him, in His Firstness. So there is nothing that is self-existent apart from Him, because He is *the All-Sufficient* (*al-ghanî*) in Himself, absolutely, and Independent of all other beings. He said: *and Allah is Self-Sufficient with respect to all the worlds* (3:97), and this is true according to both the intellect and revelation.
>
> Therefore, the existence of the world came about by Allah either for Himself or for 'other' than Himself. Because if this 'other' was Himself it would not be 'other', and also if this 'other' was Himself He would be necessarily composed in Himself and the firstness would be to this 'other'; therefore this would violate our previous statement that there is no firstness to anything with or before Allah.

So if this 'other' is not Himself, then it would be either existence or non-existence. But it is impossible to be non-existence, because non-existence can not bring the world (from its non-existence) into existence because there is no any priority to any one of them (i.e. the world and this 'other' that is non-existence) to come into existence, since both of them are non-existence, and non-existence has no effect because it is null.

On the other hand, this 'other' can not be existence, because then it either exists by itself or not (i.e. through something else). And it is impossible that this 'other' exists by itself, since it has already been proved that there can not be two self-existent beings.

So it remains that this 'other' exists by something else, and the meaning of the possibility of the world is nothing but that it exists by something else. Therefore this 'other' is the world or (some part) of the world.

Also, if the existence of the world by Allah is due to 'something' without which the world would not exist – whether this 'something' is called 'will' (*irâda*), 'wish' (*mashî'a*)', 'knowledge' (*'ilm*) or anything you want which is required by the 'possible' in order to exist – then the Real would not (be able to) do anything without this 'something'. But that implies nothing but needfulness, which is impossible for Allah, because Allah is absolutely Self-Sufficient, since He is, as He said: *Self-sufficient* (Independent) *with respect to all the worlds* (3:97). And if it is claimed that this 'something' is the Essence Himself, then we say that it is impossible for anything to be 'in need of' itself, and since He is Self-sufficient in Himself, then this (false supposition) would lead to the same contradiction – i.e. being Self-sufficient and yet needful in Himself at the same time – and all this is impossible.

Therefore, since we have already disproved the existence of any 'other' (determining cause of the existence of the world), we conclude that the existence of the world, inasmuch as it exists through something else, is (causally) related to the Necessarily Self-Existent (*wâjib al-wujûd li-nafsihi*: i.e. Allah or the Real), and that the essence (*'ayn*) of the 'possible' itself is the locus for the effect of the Necessarily Self-Existent's bringing the possible into existence. It can only be properly comprehended like this.

Hence (such intrinsic distinctions as God's) 'Wish' (*mashî'a*), 'Will' (*irâda*), 'Knowledge' (*'ilm*) and 'Ability' (*qudra*) are (all) Himself – may He be exalted far above any multiplicity within Himself! Indeed His is *absolute* Unicity, and [in the words of the famous Sura 112, *al-Ikhlâs*] He '*is the One* (*al-Wâhid*), *the Unique* (*al-Ahad*), *Allah the one on Whom all depend (al-Samad)*; *He did not give birth* – for then He would be a preceding (cause); *nor is He born* – since He would then be a result; *nor is there any "other" equivalent (kuf') to Him*' – for in that case the existence of the world would be the result of *two* preceding causes, the Real and Its 'equivalent' – may Allah be exalted (above that)! So this is how He described Himself in His Book when the Prophet, peace be upon him, was asked to describe his Lord: then He sent down the Sura *al-Ikhlâs* (just quoted here)

> (according to the circumstances described in a hadith: *Kanz*, 4734) to get rid of all (supposed) sharing (*shirk*) with other than Allah, by those high qualities and descriptions. So there is nothing Allah negates in this Sura or approves, but those negations or approbations are some people's opinions about Allah.
>
> [I.291.8]

The importance of this long paragraph comes from the fact that it shows the basis of Ibn 'Arabî's distinctive view of creation, the distinctive – and extremely controversial – view that many scholars have traditionally called 'the oneness of being (*wahdat al-wujûd*)', but which has been widely misunderstood. Ibn 'Arabî himself could not describe it plainly, simply because it is a reality whose direct perception depends not on logic, but – as Ibn 'Arabî stressed in the passage quoted at the end of section 2.1 – on a rare inspired experiential 'tasting' restricted to the spiritual elite of the 'Solitary Ones'. As soon as it is spoken, it is likely to be misunderstood. What Ibn 'Arabî tries to prove in the passages just quoted, as in many chapters of the *Futûhât* and other books, is that the existence of the world is solely dependent on the existence of the Real, Who is One and Unique – and that this ultimate dependence of the world on the Real is an essential property that accompanies everything in the world at all times. On the other hand, nothing was added to Allah through His creating the world, as for Him (in His Knowledge) the world is as if it is always there. We shall come back to this important issue of the oneness of being (especially in Chapter 5) as we continue our exploration of Ibn 'Arabî's view of time, because it is truly the key to understanding his cosmology and theology.

2.4 The origin of time?

If we can not speak about the origin of the world in time, we can still ask about the origin of time in the world. As Ibn 'Arabî pointed out, we can not ask when time began, because the word 'when' requires time to be defined beforehand. But we can ask *how* did time begin?

In response to this cosmological question, Ibn 'Arabî argues that both the natural and para-natural types of time have originated in the Universal Soul which has two forces: the active and intellective (*quwwa 'amaliyya* and *quwwa 'ilmiyya*). The active force is in charge of moving bodies and objects,[11] while the intellective force is capable of perceiving knowledge, or updating the soul's spiritual state. So physical time (i.e. that associated with physical objects) is that in which bodies keep moving to preserve their existence, and spiritual time is that in which human beings' Heart perceive knowledge from their Lord [*Ayyâm Al-Sha'n*: 6]. Physical time, therefore, is originated from the active force of the Universal Soul, while spiritual time originates from its intellective force [*Ayyâm Al-Sha'n*: 7].

On the other hand, since natural (or physical) time is a consequence of material motion in Nature, it is originated with the isotropic orb which is the first orb in Nature in which the first body was created (see section 1.4). When this isotropic orb (*al-falak al-atlas*), was created and started to move, its motion defined natural time. But because this orb is isotropic – the same in all directions – it is not possible to measure time in this orb alone, because there was nothing to compare its motion to it. And when Allah created the second orb that includes the farthest stars (galaxies) that are appropriately called fixed stars, the (apparent) motion of those stars in this orb defined the day as the complete revolution of this orb: 'When Allah caused the higher orbs to move and He created days in the first orb (the isotropic orb) and defined it (the day) in the second orb which is the orb of the apparently fixed stars' [*Ayyâm Al-Sha'n*: 6].

However, as Ibn 'Arabî says, the motion of the isotropic orb was actually determined from above; it started when the first degree of Gemini was matching the divine 'Foot' on the 'Pedestal' (*kursî*)[12] above the isotropic orb, and after one complete revolution, the first cosmic 'Day' of creation was done, and that was Sunday. Then the process of divine creation continued through Monday, Tuesday, and so on until its initial completion on Saturday – and then it started over again [II.437.34]. Because this creative 'Day' was determined from above, it was not possible to know its duration [II.437.27]. It is true that we divide this day into 24 hours, but this is a mere convention. It is actually not possible to determine the length of this day because there is nothing else (in Nature) to compare its motion to, so we use this day to measure other relative 'days' of other orbs that are below the orb of fixed stars, and also the 'days' of the spiritual (and divine) orbs that are above this physical isotropic orb too. We shall devote Chapter 3 to explaining in more detail the seven Days of the cosmic 'Week' and how they are caused.

2.5 Space-time and the speed of light

Although time does not appear to be like space, in the theory of Relativity it is treated as a real dimension just like any one of the other three dimensions of space (length, width, depth: x, y, z). In Relativity, as we explained in the preceding chapter, any point in the universe can be expressed in terms of its four-dimensional space-time co-ordinates (x, y, z, t); we do not have time alone or space alone, but a single field called space-time.[13]

Likewise, Ibn 'Arabî describes the physical universe as something that 'is confined in time and space' [I.121.22]. Furthermore, one of the most important results of Ibn 'Arabî's view of time is that he considers that we are living in 'Saturday', while the other six cosmic 'Days' from Sunday to Friday account for the creation of the world – which is now continuously being re-created by Allah (see also section 5.6) – in space. Allah creates the three-dimensional world (actually six-dimensional/directional if we consider the two directions of each dimension) in six 'Days' from Sunday to Friday, but we human beings witness only Saturday because in the other six days of the week we (along with the rest of creation) are still being created. Ibn 'Arabî insists that this divine creative process is repeated every single moment, as we shall explain in the following chapters.

However, the result of what we have just said is that time ('Saturday'), though it is special, is still just like any one of the other six Days that correspond to the six (or three) spatial dimensions. So indeed the world, for Ibn 'Arabî, is confined in those seven 'dimensions' of space-time (6 plus 1) that are similar – since all are 'days'. This is the ultimate meaning of the many verses in the Qur'an specifying that Allah created the heavens and the Earth '*in six Days*' (corresponding to space) and that '*then He mounted* [i.e. on Saturday, in time] *on the Throne*' (Qur'an 7:45, 10:3, 11:7, 25:59, 32:4, 50:38 and 57:4). This could also be easily comprehended if we recall that the actual meaning of time is reduced to the existence of the world in the present moment, not the past or the future. Thus manifest existence is confined in space and time, so both space and time refer to existence, and they have no meaning when taken by themselves, without the things or events that happen in them. This new concept will add another aspect to the theory of Relativity that considers time as one dimension of the four dimensions of space-time, especially since Ibn 'Arabî gives exciting details about how those seven Days of the cosmic Week are interconnected, as we shall see in the following two chapters.

However, there are still many obvious and hidden differences between space and time. At the beginning of chapter 59 and in the long chapter 559 (which summarizes the key contributions of each of the preceding chapters of the *Futûhât*), Ibn 'Arabî points out the similarities and differences between space and time. 'Time', he says, 'is just like space, an extension that has no (outer) limit' [I.291.6]. Then he adds:

> Space is an attribute of something that exists, but time is an attribute of something that is confined but does not necessarily exist. Space is defined by who sits in it, and time is counted by breaths. The (ontological status of) 'contingent possibility' (*imkân*) affects both time and space. Time has an (ontological) foundation that it refers back to and is based upon, which is the divine Name 'the Age' (*al-dahr*). Space emerged by the 'establishment' (*istiwâ'*) (of the All-Merciful on the Throne, 20:5), and time emerged by the 'descending down (of the Lord) to the (lowest) heaven' (referring to the hadith: 'Our Lord, may He be Praised, descends every night, in the last third of the night, to the lowest sky . . .' [*Kanz*: 3351, 3355, 3388], see also section 2.14). But there was time in the Dust (*'amâ'*) even before the 'establishment'. . . . Time is a circumstance for an event just like meanings for letters, and space is not like a circumstance, so it is not like the letter. Time is confined through division by 'now' and does not necessarily require the existence of objects, but space can not be comprehended without objects (that occupy it), so it is a kind of (ontological) 'home' (for what is created in it).
>
> [IV.337.5]

On the other hand, the concept of using time to measure distance was already used by the ancient Arabs who used to measure distance by how it took them to travel through it, usually by camel. But Ibn 'Arabî uses this concept in a more abstract way that can be compared with the form of measurement that is now widely used in astronomy: the light year. In many places he repeatedly says that the distance between this particular celestial orb and that orb is a particular number of 'years', without specifying what speed or form of motion might be involved. For example, he says that the distance between the top and bottom of Gehenna is 'seventy-five hundred years' [I.297.15]. And in other places he says that Gehenna is (or 'will be' [I.297.17]) in the entire space situated from the Earth to just below the orb of fixed stars (the constellations of the Moon mansions) [I.303.9, III.440–441].

Now according to modern astronomy, the distance from the Earth and our solar system to either extreme of the width of our Milky Way galaxy roughly equals the distance travelled by light in 7,500 years. So in effect one could argue that Ibn 'Arabî actually used the unit of a kind of 'light year' to measure distance, more than seven centuries before modern astronomers, and that he gave a very accurate value of the width of what is now known as the Milky Way galaxy.

Regarding the speed of light, Ibn 'Arabî declares that 'nothing is faster than sight (*basar*) among the (human) senses; the time of opening the sight is the time of its seeing the fixed planets (stars) or what is above them or between them despite the large distance that could not be reached for thousands of years by foot' [IV.431.34; see also I.702.20, II.402.30].

Furthermore, in chapter 8 of the *Futûhât* Ibn 'Arabî mentions many extraordinary and mysterious facts about another 'earth of Reality' (*ard al-haqîqa*) – another world existing in the *barzakh* or 'divine Imagination', and accessible to spiritual travellers – which is

> an earth so spacious that the Throne and what it includes, the Pedestal, the heavens, the earths, what is beneath the soil, all the Gardens and Gehenna, would all be just like a ring in (comparison to the vast extent of) the desert of this 'earth'.
>
> [I.126.30].

He then talks about his and other Sufis' (spiritual) visits to this Earth and that the life there is so extraordinary that many of the logically impossible things for us would normally exist there. One of the things that he mentioned about this Earth is that 'the speed of their (people's) travel on ground or by sea is faster than the perception of sight when it sees things' [I.128.26].

It is perhaps relevant to mention here that one of the consequences of modern theories of high-energy elementary particles is that each particle (such as electrons, protons and neutrons) has an anti-particle which, upon meeting with its counterpart, would annihilate and convert together into light or electromagnetic waves (Trefil 1938: 53–67). Likewise matter (atoms) has anti-matter that may exist somewhere under extreme circumstances, such as in the core of hot stars and galaxies, though it has also been made in laboratories (Schewe and Stein

1999). Some scientists say that, like matter and anti-matter, there should be an anti-universe. In this anti-universe all the laws of physics would behave strangely. For example the speed of light in our universe is a maximum terminal as we have seen above (the theory of Relativity), but in the anti-universe it would be a minimum terminal. Such a case, therefore, is predictable in Ibn 'Arabî's cosmology, as he mentions in regard to this mysterious 'earth of Reality'.

2.6 Motion

Time, therefore, is necessary to describe motion. But the answer to the question 'what is motion?' may not be as obvious as it might at first appear. Matter is in continuous motion, and objects require a cause to move; this is undisputable philosophical fact. But the basic issue in the philosophy of motion is whether the matter-in-motion can be itself the cause of its motion. The dialectical explanation considers that matter is the most primary source of the development of completion, and therefore it can be itself the cause and subject of motion. Metaphysical philosophy, on the other hand, insists on differentiating between that which moves and the mover. This is because motion is a gradual development and completion of a deficient thing, which can not by itself develop and complete gradually – and therefore can not be the cause of completion.

Osman Yahya listed a book with the name '*kitâb al-haraka*' ('the book of motion') (OY: no. 223), but as in the case of '*kitâb al-zamân*' this book is not found. However, Ibn 'Arabî talks about motion in some detail in the long chapter 198. There [II.456–458] he affirms that: 'everything in the world that moves and rests does not move and rest by itself, but by a mover and a rester', but he adds that: 'this mover either moves the object by itself or by its will to move it; so those who believe the mover moves the object by itself say that motion is "created" in the object, so motion by itself – when it is in the object – causes it to move'. And the same can be said regarding rest. 'But if the mover moves the object by its will, it will do that either by an (intermediate) means or without a means.' Then he adds that, if the mover is the object itself, then it has to have a will 'like the motion of the human being who moves under his will in the (six physical) directions'.

Ibn 'Arabî then differentiates between the regular circular motion of the orb (*al-falak*, the celestial sphere of each planetary heaven) and the motion of objects, where 'the motions of the orb are tidy and in a sequential manner like the motion of the millstone; so each part does not depart from its neighbouring (part), while the motion of (the four sub-lunar) elements is different'. Then he explains that 'the motion of the elements is entwined, where the part departs from the neighbouring part and occupies new places different from the ones it was in'. He also adds that

> the motion of the orb – for us (the Sufis) – is like the motion of the human being in the directions . . . the orb moves by (its) 'will' in order to give

> out what is (inspired) in its heaven by the divine Command which causes the things (to occur) in the (earthly) elements (*'anâsir*) and the generators (of earthly changes, the *muwallidât*); so as a result of this (orbs') motion, time emerges. So time has no effect in its (the orbs') appearance, but rather it affects (only) what is below it. Time does not affect the appearance of the orb, because it is itself the appearance (the result in the lower elemental realms) – whereas the things that happen and appear in the orbs, the heavens, and the higher world have causes other than time.
>
> [II.456–458]

We have to admit that physicists habitually accept a very naive concept of motion, usually expressed by the formula 'velocity is distance per time' ($v = s/t$), which is usually used for a simple uniform motion on a straight line though other complicated motions have more complicated equations that are all based on this simple concept of distance per time. Such a simplified concept of motion has been working nicely for many centuries and, although modern theories slightly corrected these classical (Newtonian) equations, they did not address the more philosophical question about the nature of motion itself. To answer this question, one has to verify whether space and time are discrete or continuous, an issue that (as we saw in Chapter 1) still persists and is unsettled even in the latest theories. However, we find some philosophers, like Zeno, who argued that – whether we consider this way or the other – we shall inevitably end up with some irresolvable paradoxes (see section 7.4).

Ibn 'Arabî, on the basis of his theory of the oneness of being and the principle of continual re-creation (see section 5.6), gives a clear and far more extensive definition of 'motion' (*Haraka*: Ibn 'Arabî's wider definition here reflects the fact that this Arabic philosophical term refers not just to physical motion but far more extensively to all kinds of 'change' in general) which is utterly different from the simple notion of just a distance in time. In the same chapter (198) that we just quoted above, he says:

> Then you have to know that the truth about motion and rest is that they are two states of the natural embodied (*mutahayyiz*) things . . . And that is because the embodied thing will necessarily need a volume (*hayyiz*) to occupy by itself in the time of its existence. So it may either be in the same place (*hayyiz*) in the next time, or times, which is called 'rest', or it is in the next place in the next time and in the following place in the third time. So its appearing in and occupying these places one after another can happen only by 'changing' from one place to another, and this may only be due to a cause. So it would be fine to call this change 'motion', although we know there is nothing but the embodied thing itself, the place, and the fact that it occupied a place next to that which it occupied before. But those who claim that there is some (real) thing called 'motion', which got into the embodied thing and caused it to change from one place to another, they have to prove it!
>
> [II.457.27]

With the above definition of motion, Ibn 'Arabî has in mind his basic principle of the 'ever-renewed creation', which suggests that the entire world is continuously being re-created every single moment of time, which we shall discuss in detail in Chapter 5. Therefore there is no real motion like that which we habitually perceive in the human 'common sense' or 'estimative' faculty (*wahm*); in reality there is only a 'change of place': i.e. the thing that is the subject of motion is being re-created in different places (not moved between them), so we imagine motion. At the end of his short book *Al-Durrat Al-Baydâ'* (*The White Pearl*) (*Al-Durrat Al-Baydâ'*: 142), Ibn 'Arabî wonders how (the general) people (not to mention physicists and philosophers) do not so easily realize the delusion of motion and space. He says that 'everything that moves does not move in occupied space (*malâ'*), but it moves in a void (*khalâ'*)'. Then he explains that the thing may not move into a new place until this new place is emptied. So by simple logic, this (false) assumption would lead to the conclusion that the result of an action would occur before the action itself. For example when you fill a cup with water, the air already in the cup will have to be gradually evacuated as water pours in. At any instance (the smallest duration of time), before the water (the cause) can replace the air, the air has to be displaced (the result). So the result happens before the cause. One may argue that in this case both the cause and the result could happen at the same time. This, however, is also prohibited according to Ibn 'Arabî who asserts that the entities of the world can be created only in series one after the other (see sections 5.6 and 7.3). Thus this radically different conception meticulously challenges Newton's law of action/reaction – which practically speaking always holds true, but which seems to be philosophically deceiving. So the mere concept of motion apparently violates causality, the most fundamental principle of physics, and common sense. Actually, Ibn 'Arabî (following earlier radical theories in kalam theology) even questions causality itself – as we shall see again in Chapter 7 – where he affirms that Allah says: 'I create the things next to the causes and not by them' [II.204.13]. Though this does not deny causality itself (i.e. the appearances of regular 'natural' causes), it does suggest a radically new type of *strictly divine* causality.

Ibn 'Arabî concludes, therefore, that motion is only a new creation in different neighbouring places; there is no actual 'path' of the object between its start and the destination points when taken on the smallest scale of time (i.e. when time itself is quantized). On the basis of this novel definition of motion, we shall be able to resolve Zeno's famous paradoxes that are discussed in Chapter 7. But this is also what happens, according to modern physics, in the atom where the electrons 'jump' between the energy levels (which have different distance from the nucleus) without any possible existence in between. The reason for this is that the energy of the atom is quantized, and when this energy changes, either by absorbing or emitting photons, the distance of some of its electrons from the nucleus will change correspondingly. So because the energy is quantized, this distance has to be quantized too; the electron therefore may not stay in between the orbits at all, or even smoothly jump between them; it may exist only in this orb or in the other orbit that is at a discrete distance from the first (Wehr *et al.*

1984: 72). In the Qur'an, it is also said that this is what happened to the throne of the Queen of Sheba when it refers to the unnamed man '*who has knowledge from the (divine) Book*' moving her throne from Sheba to Solomon's court '*in a blink of the eye*' (see section 7.2).

2.7 Relative and curved time

In the literature of Sufism and Islamic spirituality, we read a lot of fantastic stories that apparently look 'imaginary' even to physicists who are familiar with the theory of Relativity and the concepts such as time travel that we have seen in section 1.9. Ibn 'Arabî refers to the relativity of time in many direct and indirect ways. He explicitly says: 'minutes are years while sleeping' [IV.337.1]. But 'sleeping' here does not necessarily mean usual sleep, it could be any state of imagination or realization that momentarily isolates Sufis from witnessing the visible world while their spirit are occupied with other dimensions of being. For example, he speaks in chapter 73 of the *Futûhât* about the 300 spiritual knowers (s. *'ârif*) 'whose hearts are like the heart of Adam'. There he says that:

> If a knower (of those three hundred) is taken (to witness) one scene of the Lord's scenes (*al-mashâhid al-rubûbiyya*), they receive in one of its 'days' (i.e. 'the Lord's Day', which equals a thousand earthly years, 22:47) at that moment (when they are taken to the Lord's scene) divine knowledge (equivalent to) what others get in the world of (normal) senses in one thousand normal years with hard work and preparation. So this is how the divine knowledge that anyone from among those three hundred achieves when they are taken out of their own (carnal) soul and is confined in one Lord's Day. The person who can appreciate what we have said is only whoever has tasted that, when (normal) time was folded up (*tayy*) for them in that moment, just as distance and other quantities are folded up for the eyesight whenever someone opens their eye and looks at the orb of fixed stars: at the same time when he opens their eye, the rays (of his eyesight) are connected with the bodies of these stars. So look how big is this distance and this velocity (of our normal eyesight, in that case)!
>
> [II.9.23]

On the other hand, one of the main consequences of the modern general theory of Relativity is the curvature of space and time, and this conception is explicitly referred to by Ibn 'Arabî when he says in poetry: 'the age has curved on us and bent (*hadaba al-dahr 'alaynâ wa inhana*)' [I.202.7]. It is important to notice here that he used the word 'the age' (or eternal, 'divine time'), instead of time, because in modern cosmology the curvature of time is apparent only at very large scales, and we shall see below (section 2.19) that for Ibn 'Arabî 'the age' includes not only time but also space.

Also regarding time travel, which is widely known in science fiction and theoretically allowable in the theory of Relativity (see section 1.9), many Sufis and similar figures across many other religious traditions have of course frequently referred to their experiences of various forms of 'travel' across normal boundaries of time. Ibn 'Arabî, for example, mentions the story of al-Jawhari who went to take a bath in the Nile and when he was in the water he saw, like a vision, that he was in Baghdad and he got married and lived with his wife for six years and had children. 'And then he was returned to himself (from this momentary vision). . . . And after few months this woman, whom he saw in the vision that he had married, came looking for his house (in Egypt), and when he met her he knew her and knew the children' [II.82.22].

So according to this story, al-Jawhari travelled to another far-away place and lived there six years, all in a moment of his actual time at his first location in Egypt. Ibn 'Arabî's original readers would of course immediately connect this experience of the 'folding of time' (as the Sufis called it) with the narratives of the Mi'râj or ascension of the Prophet through and beyond all the heavens in a single night-journey (*isrâ'*).

In addition to this famous Mi'râj in which the Prophet Muhammad travelled vast distances in a very short time,[14] Ibn 'Arabî himself spoke in detail in chapter 367 of the *Futûhât* about his own numerous ascensions, although he affirms that his experiences were only spiritual, while the Prophet's ascension was both physical and spiritual [III.342.32].[15]

According to his accounts of this type of spiritual ascent in chapter 367, the physical elements of the Sufi's body dissociate and return to their corresponding natural place – earth to earth, water to water, air to air and fire to fire – and after that their spiritual self enters the celestial spheres to meet the spirits of the prophets inhabiting each sphere and to learn from them. Then one may even ascend further to the highest spiritual dimensions, as Ibn 'Arabî also describes in greater detail in his highly autobiographical *Kitâb Al-Isrâ' ilâ Al-Maqâm Al-Asrâ*.[16]

Just as is specified in the theory of Relativity, a person who undergoes time travel will encounter many more events than those who stay in their place. The big difference, however, is that Relativity anticipates that time travellers will encounter much longer (real) times, and that they will realize after they come back to their starting point that many generations have passed away. This has led to many strange paradoxes like the 'twin paradox' (D'Inverno 1992: 38). For Ibn 'Arabî, however, the issue is far more simple and realistic: the only difference between the spiritual 'time traveller' and others is that travellers will acquire much more knowledge or spiritual realization, because they encounter more events in a (outwardly) short period of time. For Ibn 'Arabî, time, after all, is imaginary, so more time means more events and more events means more knowledge. The Prophet Muhammad (and others like Ibn 'Arabî in their purely spiritual 'ascensions') encountered in the night of the ascension a multitude of events that normally need many years.[17] Likewise, in the illustrative cases of al-Jawhari and the three hundred spiritual knowers mentioned above, other people around them did not feel any noticeable change. This subject is very close to the case of sleep and dreams except that the Sufi in the ascension is the one who is awake because Ibn 'Arabî affirmed that the Prophet Muhammad said: 'people are asleep, and when they die they wake up' (this hadith is not found in standard hadith collections, but is widely quoted by Sufis and especially Ibn 'Arabî [I.313.11, II.379.33, IV.404.16]), which means that our perception of this world is like a dream and those who are 'awake' (the spiritually realized people) will experience time in a different way. So a person in deep sleep will not feel the time as those who are awake next to him or her experience it. Similarly, the student who pays attention to the teacher will acquire more knowledge than those who are absentminded.[18]

In the same way, the 'Unique (spiritual) Pole' (*al-qutb al-fard*) is called *sâhib al-waqt* ('the master of time', see chapter 336 of the *Futûhât* [III.135], and also *Al-Mu'jam Al-Sûfi*: 678–683), because he is always in full attention to what Allah wants of him every single moment. The true master of time witnesses everything in the world (since all that is a kind of reflection of his own spirit, or even part of it) all the time: that is to say, he witnesses the created world throughout space and time. This high state of attention, however, is only attainable by *Al-Qutb* ('the Pole' or 'Axis' of the spiritual universe) who is at the top of the hierarchy of the saints [II.6.28]. This Qutb, like Ibn 'Arabî himself, is a man whose heart is like the heart of the Prophet Muhammad [I.151.6], 'and the one who is on the heart of Muhammad, peace and prayers be upon him, has the corner of the Black Stone (in the Kaaba), and that is for us thanks to Allah' [I.160.24, see also section 6.2].

The spiritual 'Pole' therefore is witnessing 'out of time' – which again means that he is witnessing everything in the world, spatially and temporally. Other Friends of God (the *awliyâ'*) may attain this high state of awareness to a relative extent, 'though this is very rare amongst the (special) people of Allah, and it is (only) for a few of them, the people of attention, those who never overlook Allah's rule in things' [II.539.27]. This also explains how some divinely illuminated people, like the prophets, encountered with Allah things that would not normally occur to normal people. For example, Allah's direct speech to Moses can be explained only, according to Ibn 'Arabî, when we consider that Moses was out of time when Allah spoke to him – since otherwise Allah would be confined to time, and this is not possible. Ibn 'Arabî explained this in some details in his short *Book of Eternity*, that is one of 29 short treatises published together in the famous book known as *Rasâ'il Ibn 'Arabî*.

We actually always live in a relative time, but, although we encounter a relative number of events, time itself has no reality. Because we measure time by other standard events, like the clock's ticks or the Sun's motion, we do not feel the relativity of time. But if we measure it by our own internal activities (or what is known as psychological time), we shall always be travelling through time.

However, real travelling to the past is not possible at all, since time itself does not go back: once a form is created and goes into the past it never comes back again, although it is possible to remember past events: 'The past that has gone never comes back, but the similar (form) may come back; so when it comes back it causes (someone) through itself to remember that which was like it and has gone and is past' [II.186.27].

However, Ibn 'Arabî is well aware that it is quite possible to interact with the spirits of the dead (who are 'dead' only to this world, not in their vast spiritual realm of the *barzakh*) because they may become visibly embodied in a spiritual form in this world [I.755.10], as frequently happens in various spiritual experiences. For example, Ibn 'Arabî mentions his personal encounter with Ahmad al-Sabtî, the son of the Abbasid caliph Harûn al-Rashîd, whom he met while circumambulating the Kaaba. When Ibn 'Arabî first saw him he doubted his case, because he saw him not pushing nor being pushed, and going through between the two men without separating any space between them. So he realized he must be a visibly embodied spirit, and he went to him and spoke with him and asked him why he was called al-Sabtî (see his answer in section 3.6) [I.638.32].

Travelling to the future and meeting people who are not born yet is also possible, though it may also be a kind of simulation or personification of their spirits, and Ibn 'Arabî also mentioned similar things:

> And I have seen all the messengers and the prophets, witnessing them all by direct vision. And I talked to *Hûd*, brother of *'Âd*, in particular from their group (Hirtenstein 1999: 84–86). And I have seen all the people of faith, also by direct eye-witnessing, all those who have been among them and those who will come to be, until the Day of the Rising: the Real showed them to me on a single plane on two different occasions. And I accompanied (for spiritual learning) a group from among the Messengers, in addition to Muhammad – may God bless him. (For example), I recited the Qur'an to Abraham *al-Khalîl*; and I returned (to God) by the hands of Jesus; and Moses bestowed on me the (inspired) knowing of (spiritual) unveiling and clarification, and the knowing of (the spiritual meaning of) the alternation of '*the daytime and the night*' (2:164, 3:190, etc.). So when I had assimilated that knowing, the night-time disappeared and the daytime (*nahâr*) remained all the day long, so the Sun never (again) set for me nor did it rise – and this unveiling was a notification from Allah that I would have no part of suffering in the hereafter.
>
> [IV.77.27]

2.8 The discrete nature of time

The most important and distinctive of Ibn 'Arabî's ideas about time is that he considers it to be quantized. Thus Ibn 'Arabî declared in the *Futûhât* and other books that 'the smallest time is the single time that does not accept division' [II.384.31].

There has been a great deal of debate in the history of philosophy and science as to whether time (and space) is discrete or continuous, though most philosophers and scientists deal with time as an infinitely divisible quantity. However, the Ash'arite theologians' distinctive physical theory (of the *jawhar*, or 'indivisible atom') is entirely built on the discreteness of space and time, and Ibn 'Arabî himself acknowledges his debt to them for this understanding. For example, al-Bâqillânî, one of the famous Ash'arite theologians, suggested an atomic nature of time, according to which in each 'atom' of time the entire world annihilates and is re-created in a slightly different form (MacDonald 1927: 326–344). This perspective is at least verbally and conceptually very much in accord with Ibn 'Arabî's fundamental principle of the ongoing re-creation of all things (see section 5.6) – although the Ash'arite theologians did not make any reference to the key experiential basis of this spiritual insight which is so central to Ibn 'Arabî's discussion of the 'ever-renewed creation'. What is also new and distinctive here about Ibn 'Arabî's understanding of this conception of time is that he argues that the actual 'quantum' of time equals our normal earthly day itself [*Ayyâm Al-Sha'n*: 6]. It is not easy to bridge the gap between this metaphysical hypothesis and our everyday familiar experience of indefinitely divisible time: of the year into months, the month into weeks, the week into days, the day into hours, hour into minutes, minutes into seconds, and so on apparently infinitely. Ibn 'Arabî, however, explains plainly why time has to be discrete according to his understanding (or 'according to his Day'), and we shall devote Chapter 4 to discussing this difficult issue in more detail. Here we can give only a general preview.

For Ibn 'Arabî, time itself does not have a separate existence, but is reduced to motion or, more precisely, to the ongoing creative acts of God, or cosmic 'events' (each discrete divine 'Task' or *sha'n*) of each 'Day'. And according to the Qur'anic description of God, *each Day He is upon some (one single) task* (*sha'n*) (55:29). And since, as Ibn 'Arabî explains [*Ayyâm Al-Sha'n*: 10], Allah specified in this verse that He is every Day in '*one*' task, and not many as we perceive in our illusion, which witnesses a multitude of events every day because of the intertwining between these Divine Days and our normal days. This means that this 'Day' has to be indivisible, because only one divine action or event should be happening in it [*Ayyâm Al-Sha'n*: 6]. As we have seen, Ibn 'Arabî uses the same Qur'anic term and therefore calls this fundamental quantum of time 'the Day of Task' or divine event (*yawm al-sha'n*), or – using an expression taken from the physical theory of Ash'arite theology – 'the singular (unique) time' (*al-zaman al-fard*) [I.292.16, II.82.6]. Therefore the single 'Day of task' in reality equals our normal day; or more precisely, a full revolution of the celestial sphere as viewed from the Earth. Ibn 'Arabî helps to clarify this counter-intuitive understanding of the foundational divine 'Event' by introducing some related new concepts that are also based on certain Qur'anic verses.

The first related concept is that the world is surely re-created every singular day over and over again [II.208.26, II.385.4]. So we (our souls), as part of this world, live or ordinarily experience only our very limited portion from this singular day; a single instantaneous 'moment' of time that equals the global '24 hours' taking place within the entire world at that particular instant, but as divided up (in actual perception) by the total number of perceiving entities and their perceptions in the world. During this divine creative 'glance' (*lamha*), we perceive a still picture of our limited perception of the world, after which we intrinsically cease to exist. Then in the second 'Day' (actually in the second 'Week'!), we live another moment to perceive a different still picture, owing to the ever-new creation created by Allah in this singular day. Thus, through a succession of these instantaneously re-created new 'pictures' of the whole, we observe what appears to be motion, just as with the illusion of cinematic projection. We said 'in the second Week' because Ibn 'Arabî showed that we are indeed only living in 'Saturday', since the other 'Days' (of Creation) of the Divine Week, from Sunday to Friday of the Days of tasks, are for space not for time, because at every moment in our life the world is re-created in six Days (for space), and it is then displayed on Saturday, but we do not witness the process of creating the world: we only witness it as created. So our life as a time is a collection of Saturdays, and each moment that we live is indeed a Week; six Days for creating the world in space, and Saturday for displaying it in time. We shall come to this important and novel concept in section 3.5.

The other related concept is that the moments that we feel flowing as day-times and night-times are actually a *collection* (of discrete time-space quanta), and not a straightforward combination of the actual flow of the single divine 'Days'. The normal days that we encounter are 'intertwined' (*mutawâlija*, v. *yûliju*) – another key Qur'anic expression – with the actually existing cosmic 'Days of tasks' in a special way that we shall explore in Chapter 4. As a result of this intertwining, we see the appearance of a multitude number of events in our normal days.

In addition to that, Ibn 'Arabî also reminds us that at every moment there is a full day around the globe: evening somewhere, morning somewhere else, and noon in other places [*Ayyâm Al-Sha'n*: 6]. Therefore, at every *moment* for us, which is a full day when viewed globally, Allah creates a single event in the world; and then He re-creates the world in ever-new events at each succeeding moment. The day that we perceive and experience is therefore a collection of successive 'snapshots' of the actual 'Days of events' which are the actual existing basis for our experience of the flow of time.

As we said above, the idea of discrete time (and space) is not new in the history of philosophy and physics, though it had been completely discarded after the advent of classical Newtonian mechanics. However, many philosophers (such as Kant, Russell and Leibniz) have opposed Newton's hypothesis that space and time have a separate linear and continuous entity, as we showed in section 1.8. With the advent of Quantum Mechanics, Field Theory, the Theory of Quantum Gravity and the Superstrings Theory, the idea of a quantized time (and space) was revived again, and much work has recently been done in this respect as we mentioned in section 1.9.

The human mind naturally thinks of quantities as either discrete or continuous; there is no other way. A closer examination of Ibn 'Arabî's view of time, however, shows that it is indeed neither discrete nor continuous. We must remember that he considers time as imaginary after all, as well as most other quantities such as space and even mass (see section 7.9). As we indicated previously, such seemingly strange conclusions result directly from Ibn 'Arabî's fundamental theory of the oneness of being: i.e. that the apparent 'parts' of existence are merely manifestations of *a single real existence* that is One and Unique, neither multiple nor divisible. The notion of either discreteness or continuousness is indispensable when we imagine multitudes, but with absolute Unity there would be no meaning to such conceptions. Therefore, Ibn 'Arabî's concept of time is that it would be discrete if we approach it on the (ultimately imaginary) plane of apparent multiplicity, but in reality there is no such reality as 'time' at all. The same perspective can be applied to space.

2.9 The 'chest' of time

Ibn 'Arabî asserts that

> You should know that everything has a 'chest' (*sadr*), and knowing it in this path (of mysticism) is one of the noble sciences and knowledge. This is because the world and every kind (in it) is created according to the image of the Human Being (*insân*), who is the last existent, and the Human Being alone is created according to the divine Image, both its inner and outer dimensions, and Allah created them with a chest. So between the Real, to Whom is the firstness, and the Human Being, to whom is the lastness, there are 'chests' whose number no one knows other than Allah.
>
> [II.652.23]

The 'chest' of a thing is its front, the first thing to appear, which is why it is so important to know it. Ibn 'Arabî here applies and generalizes the verse in the Qur'an that says: '*And whomsoever Allah wants to guide, He expands his chest for "surrender" (islâm), and whomsoever He wants to leave astray, He makes his chest narrow and constricted as if he were engaged climbing up into the sky*' (6:125).

But Ibn 'Arabî adds that 'the chest is in the second state of each form' [II.652.27], since the first state is the essence (the 'heart') of the form itself and the chest is *its first appearance or effect*. Then he lists many things and their corresponding 'chests', and he says that 'the chest of time is the time (instant) of the Dust's acceptance of the form' and 'the chest of days is Monday' [II.652.27].

Thus Ibn 'Arabî adds that:

> For every 'chest' there is a 'heart' (*qalb*), so as far as the heart remains in the chest, the person will be blind because the chest is a veil over him. So when Allah wants to make him or her truly seeing, he goes out of his chest and sees. So the causes are the 'chests' of the existing things, and the existing things are like the hearts; as far as the existing (thing or person) is looking at the cause that it originated from, they will be blinded from witnessing Allah Who (actually, ultimately) created them. Thus when Allah wants to make (the person) truly seeing, they stop (or He causes them to

> stop) looking at the cause that Allah created them 'with' (*'indahu*: i.e. as opposed to 'through', *bihi*) and looks (instead) at the 'special Face' that is (uniquely) between their Lord and them.
>
> [II.652.35]

We are, therefore, normally 'imprisoned' in the chest of time, just as the heart of the person lacking faith is imprisoned in or 'veiled by' his or her chest. To overcome time (and space), we therefore have to break out and perceive the reality of existence through the 'heart', and in particular the special inner relation or aspect between the 'heart' and the Real.

2.10 Circular time and cyclical time

Ibn 'Arabî repeatedly describes time as a 'circle' (*dâ'ira*), [I.387.33], which does not have a beginning or an end, but when we specify a point on this circle (the present; the point in time in which we exist now) and look in one direction, whether to the past or future, we do set a relative beginning and an end. So the present (now) joins together the two ends of time in a circle [I.387.32, III.546.30].

It is very difficult to imagine this, as it was very difficult to convince people that the Earth is spherical when this idea was first introduced, precisely because most everyday activities show us only a small portion of the Earth's surface which appears flat to us. Yet time is not like space. In the case of the Earth it was relatively easy to prove that it is round because we can view its curvature – from space, or from a great height – all at once. But the problem with time is that we can normally witness only the present moment of it, not the future nor the past; we can only imagine them. Therefore, in order to understand the meaning of 'circular' time, we have to imagine that the whole of all existence (what we perceive as future, present and past) exists all at once. This whole existence is then like a circle: i.e. a curve that does not have a visible beginning or an end when we look at it from outside. When we sit on the circumference of this circle and look in one direction, we set a beginning and an end. In the same way: the present moment in which we exist is a point on the circle of the whole existence, this point defines the future and the past and it also defines an imaginary beginning and an imaginary end of time: imaginary because the whole circle of existence (of 'the Age', *al-dahr*) is infinite. The imaginary beginning is the eternity *a parte ante* (*al-azal*) and the imaginary end is the eternity *a parte post* (*al-abad*) [IV.266.3].

Circular time has yet another important meaning which is not possible to explain fully at this point, because it needs additional premises that we shall discuss in the following chapters, so we can refer to it only briefly here. As described above, Ibn 'Arabî views the world as being continuously re-created, and time is reduced to the present moment because the past and the future are only imaginary. Therefore time, or the present moment, goes in ever repeated circular motion with the re-creation of the world. In other words, the presence, which is time, goes round the world continuously and repeatedly to create it and re-create it again; so it is circular (and cyclic) in this distinctly metaphysical respect.

Moreover, this cosmic notion of 'circular time' is quite different from 'cyclical time', and the two ideas should not be confused in Ibn 'Arabî's writings. 'Cyclical' or periodic (*dawrî*) time, such as the day, the week, the month and the year, is a duration of time in which the same kind of events should be happening in the different repeating cycles of time. For example, the Sun sets every evening to start a new day (because the 'day' for the Arabs was considered to start from sunset, not from sunrise), and the Moon is born approximately every four weeks to start a new lunar month. We should again note, however, that in reality, according to Ibn 'Arabî, there is never any repetition at all [II.432.12, III.282.21]. Those cycles of time are 'similar' to each other but never identical, as the terms 'period' or 'cycle' mean in modern physics. For Ibn 'Arabî, the cycles of time are similar because they are ruled by the same divine Names, which is why we expect to see similar events. But the reason why we do not see identical events or true repetition is because of the interaction between different cycles of different divine Names [III.201.14]. For example he says that 'the motion of Sunday appeared from the (divine) attribute "the All-Hearing" (*al-sam'*) . . ., and the motion of Monday appeared from the (divine) attribute "the Living" (*al-hayât*)' [II.438.9].

Therefore the cycle of Sunday is different from the cycle of Monday, although they are both days in which the Sun rises and sets in the same way. On the other hand, and on the basis of the same statement above, Sunday from this week should be identical with Sunday from the previous week, because they both appeared from the same divine Attribute. But they are not, because they do not have the same position in the month or the year, or in other cycles that are ruled by other divine Names and Attributes, so there is never any repetition. Ibn 'Arabî nicely refers to this fact in his prayers *Al-Salawât Al-Faydiyya*, a short text found at the end of *Tawajjuhât Al-Hurûf* (Cairo: Maktabat al-Qâhira, n.d.), when he says: '(He, i.e. the Prophet Muhammad, is) the Dust (*hayûlâ*) of forms which does not manifest in one form to any two (persons), nor in any one form to anyone twice' [also in I.679.7, II.77.27, II.616.3].

We shall devote Chapter 3 to the significance of the 'Week' in Ibn 'Arabî's view, where we shall see that the week is the primary *cycle* of time, not the day. We shall also see that Ibn 'Arabî's fundamental notion of 'no repetition' forms the basis of a unique view of time and the cosmos, which will be discussed in the following chapters. However, if we accept the approximation that different cycles of time are defined by similar events, we then find that Ibn 'Arabî defines many important cycles other than the usual day, week, month, the seasons and the year. In general, every divine Name of Allah has Its own 'day' (cycle), which is a cycle of time that has a corresponding daytime and a night in the world below it (see Table 2.1 below).

A correlation can also be made here between Ibn 'Arabî's view of cyclic time and astrology. Knowing these cycles and their lengths and the specific Names that rule them may give us some insight into what kind of events may happen in

Table 2.1 Days of some orbs and divine names

The day of . . .	*Its length in terms of normal Earth days*
The Moon	28 days
The Sun	360 days = 1 year
Mercury	~30 years
al-rabb (the Lord)	1,000 years
al-mithl (the Like)	7,000 years
al-ma'ârij (the ascending ways)	50,000 years
Zodiac	12,000 years
Longest planet (star) day	36,000 years

Note
The numbers have been collected from different books of Ibn 'Arabî. We should add that – as noted in our table here – Ibn 'Arabî actually considers the year to be 360 days rather than 365.25 days; see also section 3.2 for more details

the future, but it is not easy to tell exactly what is going to happen. The issue is more like weather forecasting; by studying all the parameters we can tell with a good probability how the weather may be in the near future.

2.11 The two cycles of life

According to the well-known hadith 'Time has returned the same as it was when Allah created the heavens and the Earth' [*Kanz*: 12357], Ibn 'Arabî argues that the cosmos has completed one cycle from the beginning of creation until the point where the spirit of the Prophet Muhammad is attached to his body, and he notes that this cycle is 78,000 years. As he explains:

> Time started in Libra for spiritual justice by the name the Hidden (*al-bâtin*) of Muhammad – may God bless him – as he said: 'I was prophet when Adam was still between water and clay' [*Kanz*: 2017]. Then it (time) circled around after one (full) time cycle, which is 78,000 years, to start another cycle of time by the name 'the Manifest' (*al-zâhir*) when the body of Muhammad – may God bless him – appeared.
>
> [I.146.18]

In the beginning of the *Futûhât* [I.121–124], Ibn 'Arabî also states clearly the *order* of creation in time during this cosmic time cycle, for major events of the kinds created by Allah. The numbers that he gives clearly do not agree with modern cosmology and geology, and can not be explained or understood in those terms (as we showed in section 2.3 above).

2.12 'Days'

So we have seen in this chapter that time is in fact imaginary and exists only as the Creator's 'Days' and indeed indivisible moments, while all other divisions of

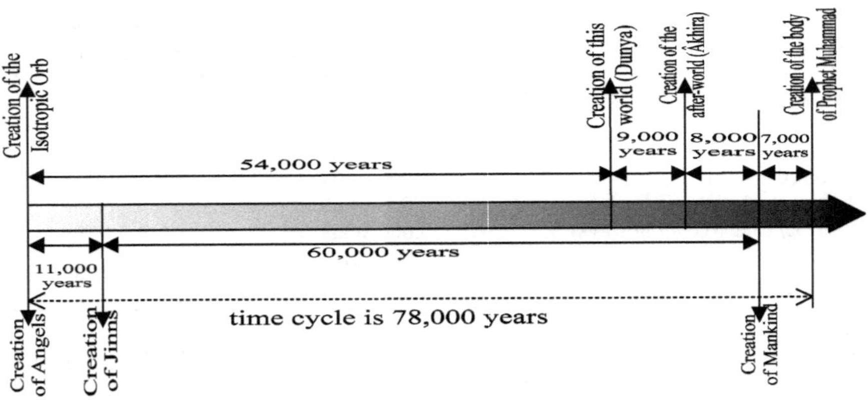

Figure 2.1 The Cycle of Life.

Note
The number of years are extracted from chapter 7 of the *Futûhât* [I.121–126].

time are conventional. For this reason we find that in the Qur'an the word 'time' is never used, while a great deal of attention is given there to mentioning different sorts of 'daytimes' and 'nights' and the relations between them. Likewise, we find that Ibn 'Arabî pays considerable attention to describing the actual meaning of the days and the relation between their different types.

As usual, the terms 'daytime' (*nahâr*) and 'night' (*layl*) are used by Ibn 'Arabî to measure time, where the daytime extends from sunrise to sunset, while the night (*layl*) is from sunset to sunrise: both of those together – always *beginning* with the night-time – are called a 'day' (*yawm*), which is conventionally divided into 24 hours. We must notice, however, that the above conventional definitions are only approximate for our practical use here on Earth, since considerable differences appear as soon as we begin to measure the length of the day, for example, with relation to the Sun or to distant stars.

In the following passage Ibn 'Arabî gives a precise definition of various units of time such as the day, the daytime, the night, the month and the year. Some of these definitions are, however, simply for practical use for the purpose of determining prayer times. More accurate variations will be discussed below and in section 3.2. At the very beginning of the extremely long chapter 69, in which Ibn 'Arabî talks about 'the secret meanings of prayer (*salât*)', he devoted a full section to detailed explanations about timing. For the purpose of demonstration and simplification, Ibn 'Arabî employs a hypothetical observer who is considered as a frame of reference. He says there:

> 'Timing' (*al-waqt*) is an expression for (our approximate) estimation concerning a thing that does not accept the actual reality (*'ayn*) of what is estimated. So it is an (approximate) supposition, just like we suppose and estimate a beginning, middle and an end in a spherical shape which in itself

and in its actual reality does not accept a beginning, middle or end. So we construe all of this in it (only) by what we construe through the effect of our supposition and estimation concerning it.

Likewise, timing (*waqt*) is an estimated supposition with regard to time (*al-zamân*), since time is circular, as Allah created it in its beginning, so it is like a circle. The Prophet Muhammad, may Allah have peace and mercy upon him, said: '*time has circulated* [i.e. has come back to the same point] *like its form was on the day when Allah created it*' [*Kanz*: 12357]. So he mentioned that Allah created it circular (see section 2.10), and timings are (only) estimated with regard to it.

So when Allah created the Isotropic Orb and it moved, the 'day' was not specifically determined and no (temporally defining) reality had (yet) appeared in (that Orb). It was like the water of the pitcher when it is (still) in the river, before it comes into the pitcher. Then when the 12 (equal zodiacal regions) were supposed in it (i.e. in the Isotropic Orb, and subsequently appeared in the second orb of the fixed stars, which has in it the zodiacal constellations), it was given specific timings, and He called them the '(zodiacal) constellations' in that (isotropic) orb – which is Allah's saying: '*by the heaven*' – (swearing by it) because of its loftiness above us – '*that has the constellations (burûj)*' (85:1). So they are these suppositions for timing.

So (when) a person stands (on the Earth) and this orb rotates around him, and this person has been given sight to look at these (spatial) suppositions, through the distinguishing signs (of the zodiacal constellations) that were determined in (the outermost sphere), then (that person) can distinguish some of its parts from others, by these distinguishing signs that are made to be references pointing to it. So (this person) fixes their eye on one supposed (area) of it, I mean on the distinguishing sign (of this or that constellation), and then the (zodiacal) orb rotates with this supposed distinguishing sign that this observer has fixed their eye on, until it disappears (from his sight). This continues, as long as they continues standing in their place, until (eventually) this sign comes back to them (in the same position). Then at that point they know that this (zodiacal) orb had completed one cycle with respect to this observer – not with respect to the orb itself (because the orbs' real motion takes much longer time than a day, see section 1.4). Then we called this cycle a 'day' (*yawm*, defined here according to the far-away sphere of the zodiacal stars and not according to the Sun, which is the day known in astronomy as the 'sidereal day').

Then after that, Allah created in the fourth Heaven (celestial sphere) of the seven Heavens a lighted planet that has a huge body, and it was called in the Arabic tongue '*Shams*' [i.e. the Sun; but the Arabs used to call both the planets (*kawâkib*) and the stars (*nujûm*) *kawâkib*, 'planets' (s. *kawkab*), but Ibn 'Arabî clearly distinguishes between them]. Then it rose in the sight (of this observer) from behind the veil (or horizon) of the Earth where this observer stands, so they called this place of rising 'the shining-place' (or 'east/orient': *mashriq*), and they called the rising a 'shining-forth' (*shurûq*), because this bright planet rose up from it and lighted up the atmosphere where this observer is.

So the sight of this observer kept following the motion of that planet (the Sun) until it was opposite them (in the middle of the sky), so he called this (state of) opposition 'the meridian' (*al-istiwâ'*). Then the planet began to descend from its meridian with respect to this observer, seeking the right side of them – not with respect to the planet itself. So they called the beginning of this descending from its meridian a 'decline' (*zawâl*) and disposition (*dulûk*). Then the sight of this observer kept following it until the body of this planet went down, so they called its going down 'setting' (*ghurûb*), and they called the place where their sight saw that it went down its 'setting-place' (or 'west': *maghrib*).

Then the atmosphere became dark for him, so he called the duration of the lightning of the atmosphere, from the rising of this planet till its setting, a 'daytime' (*nahâr*). (This name is) derived from '*al-nahr*' (the river), because the spreading out of the light in (that daytime) is like the spreading out of water in the bed of the river.

So this observer remained in the dark until that planet that is called the Sun (again) rose from the place that they called the orient, in the sight of this observer – (but) from another (different) place, close to this place that it rose from yesterday, (by a distance) which is called a 'degree' (*daraja*). So they called the duration of the darkness in which they were from the time of the setting of the Sun till its rising a night (*layl*). So the day (*yawm*, the conventional rotational day and not the sidereal day) is the sum of the daytime (*nahâr*) and the night (*layl*). And he called the positions where this planet rises everyday 'degrees' (*darajât*).

Then they saw that this bright planet, which is called the Sun, moves between those estimated suppositions (marked by the different zodiacal signs) in the (Isotropic) circumferential orb, one degree after another, until it cuts through that (first supposed position) through these risings called days, such that when it completes cutting through one supposed (position), it starts cutting through another supposition, until it completes (going through all) the 12 suppositions by cutting (them). Then it starts another cycle by cutting through these supposed positions (again). So they called (the time) from the beginning of cutting each supposed position till the end of cutting that (particular zodiacal) supposed position a 'month' (*shahr*); and they called (the Sun's) cutting through all those (12 zodiacal) suppositions a 'year' (*sana*).

Thus it has become clear to you that the night, daytime, day, month and year, are called 'timings' (*awqât*), and (also) it gets shorter till what is called hours and less – that all that does not have (real) existence in its essence, but that they are only relations and relative connections (*nisab*/*idâfat*). But what is (actually) existing is (only) the essence of the orb and the planet, not the essence of the timing and time, since they, I mean the times, are only suppositions within it. So thus you see now that 'time' is (only) an expression for something (humanly) imagined, in which these 'timings' are only supposed.

[I.387.30–388]

So the day (*yawm*) for Ibn 'Arabî is like our usual day: i.e. the full revolution of the heavens as we see it from the Earth, which is conventionally measured according to the motion of the Sun. This definition of the day works perfectly for practical issues, such as determining prayer times. But if we want to be more accurate, the day indeed is the full revolution of the orb of the fixed stars [*Ayyâm Al-Sha'n*: 6] – which is in reality (i.e. as we know today), a single full cycle of the motion of the Earth around itself with relation to far-away stars, *not* with relation to the Sun. That is why Ibn 'Arabî affirms that the 'day' (*yawm*) actually existed even before the creation of the seven planets including the Sun and the Earth, while the earthly daytime and night-time (*nahâr* and *layl*) were defined only after the creation of the Earth and the Sun. He says that

> When Allah caused these higher orbs to rotate, He created 'days' in the first orb (that is the isotropic orb, because it is the first orb to be created in Nature) and defined it in relation to the second orb (that is the orb of fixed-stars or the zodiacal constellations) which has the apparently fixed planets (stars). . . . Then He created also the Sun, so the daytime and night are caused by the creation of the Sun (that appears) in the day. But the 'day' [i.e. the sidereal day; defined by the rotation of the highest sphere(s)] existed before (the Sun's creation) . . . so when the orb of the zodiac rotates one cycle, it is called the 'day' in which '*Allah created the heavens and the Earth* (*in six days*)'.
>
> [I.140.30]

As we also showed in the previous chapter, Ibn 'Arabî showed on many occasions that all the stars are moving at very high speeds. He also showed that the stars that form the zodiac signs are, like other stars, very far away, which is why we do not realize their motion. So practically we consider these stars as fixed and therefore as a reference, but in fact the reference should be the Isotropic Orb, because it is the one that encompasses all other (material) orbs. However, because this orb has no any distinguishing sign, it can not be used as a reference. Therefore, to be more accurate, we have to measure the day not relative to the Sun but relative to stars, the constellations 'from *Nath* to *Nath*, from *Butayn* to *Butayn* or from *Thurayya* to *Thurayya*'[19] [*Ayyâm Al-Sha'n*: 6] – since it is not possible to measure it relative to the Isotropic Orb which does not have any distinguishing feature. In astronomy, this is called the 'sidereal day', which is about four minutes longer than the normal (rotational) day. The difference is due to the Earth's rotation around the Sun at the same time it spins around its axis, which causes the sidereal day to become slightly longer. Although Ibn 'Arabî accepts the usual concept of the day that is our normal day (from sunrise to sunrise) for daily needs, such as knowing the time of prayers [I.388.14], he clearly distinguishes between the sidereal day and the normal day when it comes to critical issues such as the 'intertwined days' and the 'taken-out days' that we shall explain in Chapter 4.

The real meaning of 'day' comes from the fact that in this day Allah creates the whole manifest world – i.e. the whole 360 degrees of the orb or the outermost celestial sphere, the 'Pedestal' – in it. This does not at all contradict the many verses in the Qur'an and other holy Books stating that Allah created the Heavens and the Earth '*in six days*' ('*and then*' – on the seventh day – '*He mounted on the Throne*') (see the Qur'an: 7:54, 10:3, 11:7, 25:59, 32:4, 50:38 and 57:4) because we only witness the last day ('Saturday', *al-sabt*) out of these seven days, while the other six days of creation are actually included in it as space (see section 3.6). Therefore, unlike some other Muslim theologians, Ibn 'Arabî does not find any difficulty in explaining those verses in the Qur'an that talk about Allah's creating the Heavens and the Earth in six days. Most religious scholars suggested that Allah meant 'assumed days', such that if days had actually existed then, then the time of this creation would have been six days, because they could not conceive of days before the creation of the Sun and the Earth. But Ibn 'Arabî affirms that the creation of the Sun only divided the day (*yawm*) into daytime and night-time. So he gives a dramatically different cosmological meaning to the process of creation in a 'Week' (six days plus Saturday) – as the creation of space-time at every moment – and not the commonly understood meaning that it took Allah a current earthly week to finish the creation.

However, as already noted, Ibn 'Arabî did observe that there was a difference between the Arabs and some non-Arab (*'ajam*) groups in their conventional definitions of the 'day', in that the Arabs considered the day to extend from sunset to sunset, while others considered it to extend from sunrise to sunrise. So for the Arabs, the night precedes daytime, while for non-Arabs it is the reverse. This matter has no effect on the length of the whole day itself, but its implications do have an effect on the actual unit of day and especially on its spiritual and symbolic meanings, because

> For the Arabs and the Arabic timing, it has been traditionally agreed that the night precedes daytime, since originally the Creator of time, Allah the Exalted, says: '*and a token unto them is night; We strip the day out of it . . .*' (36:37). So He made the night as the origin and took the day out of it, just as the skin is stripped off the sheep. So the (initial) appearance is to the night, and the day was hidden in it, just as the skin of the sheep appears and covers the sheep until it is stripped off. So the witnessed world (*'âlam al-shahâda*) was stripped off the unseen realm (*al-ghayb*), and our existence was stripped off the non-existence. So the knowledge of the Arabs advanced that of the non-Arabs, because the (i.e. the non-Arab) calculations are solar-based: they consider that the daytime precedes the night, and they have some right (to maintain that) in this (same Qur'anic) verse, which continues . . . '*then they are in darkness*', for 'then' here refers to the present time or the future time, and the thing will not be in darkness until the coming of the night, in this verse. So (from their perspective), the daytime was like a cover on the night and then it was taken out or removed, so '*they are in darkness*'; so the night appeared which causes darkness, so the people are in darkness.
>
> [I.716.9, also in *Ayyâm Al-Sha'n*, 7]

2.13 Days of other orbs and divine names

Ibn 'Arabî then extends the meaning of the normal 'day' as described above to the spheres of all planets (and stars) and even to symbolic 'spheres' (i.e. the orbs of spirits and divine Names), where he calls the period – i.e. full revolution of each particular orb – the 'day' that corresponds to this specific orb. In this way there are shorter days and longer days, depending on the relevant orb:

> and when Allah caused the isotropic orb to rotate . . . and made its full cycle a complete day (i.e. the normal day) without daytime and night . . . So 'Days' are different: some Days are a half-cycle, some Days a full cycle, some Days 28 cycles (days); and some are more than that, (all the way up) until the 'Day of ascending ways' (*yawm dhû-al-ma'ârij*, of *50,000 years*, described in the Qur'an, 70:4), or less than that (all the way down) until the 'Day of event'; so the degrees of Days change between these two (extremes of the) Days.
>
> [III.433.35]

There is no maximum limit to the 'Day' that one can count [I.292.17], but there is a minimum limit. The maximum limit is the Age (*al-dahr*), which is one unique day that does not repeat and has no daytime and night [III.202.5]. But this Age is infinite, whereas the smallest Day is the 'singular day' (*al-yawm al-fard*) or the 'Day of event' (*yawm al-sha'n*) [I.292.16], which is that *Day* in which Allah *is upon one task* (55:29). From the meaning of the 'singular day' as the time in which Allah creates by one single act every entity in the world, we can generalize to all other days: the 'Day' of every orb is the time in which that orb affects every entity in the world (i.e. by making a full revolution).

But again we should not confuse – as we have seen above – the revolution of a planet (such as the Earth) around its axis in respect to its Sun, and in respect to distant stars (which are appropriately considered fixed). In astronomy, the first revolution is called a 'rotational' day, while the other is called the 'sidereal' day. The sidereal day is the time the Earth takes to rotate 360 degrees, in relation to distant stars. This equals, in modern calculations, 23 hours, 56 minutes, 4.09 seconds, while the mean solar day is 24 hours. The Earth actually rotates 360.98 degrees in 24 hours; the difference is caused by the Earth's orbital motion (around the Sun). Similarly all other planets or orbs have their own respective sidereal and orbital 'days'. Also, the Moon completes a cycle around the Earth once every 27.3 days, with reference to distant stars, so this is the *sidereal* lunar month. The normal lunar month used in Islamic calendar calculations, though, is the time interval between new moons as observed from Earth, which equals 29.5 days; this is called the 'synodic' lunar month. Therefore, the Moon's 'Day', if calculated by Ibn 'Arabî's approach to the Earth's sidereal day, should be the sidereal lunar month. But we shall see in section 3.2 that Ibn 'Arabî insists that the lunar month equals 28 days exactly.

Ibn 'Arabî made it clear in several places in his *Futûhât* [I.141.17, III.549.3] and other books that the fixed stars are not actually fixed at all, but that we can not notice their motion from Earth owing to the large distance, because our age is too short to notice their motion despite their very high speed (an observation well confirmed by modern cosmology). Therefore when we say 'the day is the period of motion of fixed stars', we now know today that this apparent motion is caused by the motion of the Earth and not the stars, but since we move with the Earth, we think that the stars are moving. Indeed the actual motions of the orbs of stars have much longer periods, as is well known in modern astronomy. Actually, Ibn 'Arabî, following the accepted cosmological theories of his day, differentiates between two kinds of motion of the celestial orbs: natural or intrinsic motion, and forced or extrinsic motion:

> The smallest Day is that which we count as the motion of the circumferential orb in whose day (*yawm*) the night (*layl*) and daytime (*nahâr*) appear. So that is the shortest day for the Arabs (i.e. in their language), and it corresponds to the largest orb, because it rules everything inside it. The motion of everything inside this orb in the daytime and night is a forced motion by this orb, through which it forces (a movement of) all the orbs that it surrounds. And each one of these orbs also has a natural (inherent) motion. So every orb below the surrounding orb has two motions at the same time: a natural motion and a forced motion. And each natural motion in every orb has a specific day which is measured in terms of the days of the surrounding orb.
>
> [I.121.25]

So, for example – according to modern astronomy – the Moon naturally rotates around the Earth in about 28 days while at the same time it is forced to move with the Earth around the Sun. Likewise with most other orbs of the planets, in terms of the astronomy of Ibn 'Arabî's time: the intrinsic motion of those orbs defines a specific 'day' for every one, which equals the time needed to make full revolution around the Earth. In general, however, the day is the revolution of the specific orb, so it may have many possible lengths or measurements depending on the reference point (i.e. whether it is observed from the Earth or from other orbs or planets), but here we only consider the days of the orbs with relation to the Earth.

The relation between Days and orbs is that – in general – the longer Day corresponds to the larger orb. But this is not true for the isotropic orb, which is the first and largest (material) orb. The Day of this orb is the smallest Day, which is the day that we count, i.e. 24 hours. This is because the motion of this orb is a natural motion, whereas the motions of other orbs are a combination of this first natural motion (which is a forced motion on those other orbs) and another natural motion which is intrinsic to every orb. But if we consider the motion of the isotropic orb as expressing in reality the motion of the Earth around itself, as in modern astronomy, we can then generalize and say that the larger the orb, the bigger its Day. However, still another factor that affects the length of the Day is the speed of motion of the relevant orb. Therefore, the day that we count – i.e. the 24 hours – is actually the shortest Day, though Ibn 'Arabî sometimes mentions that a day could be a half cycle of the prime cycle of the Isotropic Orb, which is the 24-hour day [III.434.2], which is equal to the daytime (*nahâr*), but otherwise the day itself as the revolution of the Isotropic Orb is the smallest day. Then comes the Day of the Moon, which is 28 Earth days (which is a little less than the actual day of the Moon that we call the lunar month; we shall give a more detailed discussion of this point in section 3.2), and so on [II.441.29].

As in the case of the material orbs of stars and planets, Ibn 'Arabî adds that every spirit and every divine Name of Allah has its own 'Day'. That is because every spirit and divine Name has effects on the world beneath it. So when they complete a full revolution around the world (i.e. without material motion but with regard to their effects on every entity), this is their 'Day'.

On the basis of particular references in the Qur'an and Hadith, Ibn 'Arabî assigns Days with specific length to some divine Names. In addition to 'the Lord's day' which equals '*one thousand years of what you count*' (32:5), and 'the day of the ascending ways' (*dhû al-ma'ârij*), which equals fifty thousand years according to the Qur'an (70:4), Ibn 'Arabî also mentions 'the day of *Mithl*' ('the Like', i.e. the cosmic 'Likeness' of the divine, referring to the famous Qur'anic verse 42:11 – the 'Perfect Human Being' (*al-insân al-kâmil*), who is created, according to a famous hadith, 'on the Image of the All-Merciful', see *SPK*: 27–30: 276) which equals seven thousand years [*Ayyâm Al-Sha'n*: 18], and another day which equals 6,454.54545 years, though he does not mention the relevant Name [I.121.23]. He also mentions yet another day which equals three thousand years, but says that he does not know the corresponding Name [III.238.13]. All this is summarized in Table 2.1 which shows the length of some other days, in addition to those just mentioned above.

2.14 The daytime and night-time

Ibn 'Arabî not only extends the concept of the known day to the orbs and divine Names, as we have seen above, but also he gives a very broad meaning of daytime and night (*nahâr* and *layl*), suggesting that every orb and divine Name has a corresponding daytime and night like our normal daytime and night. Just as our daytime and night are caused by the apparent motion of the Sun, these other infinitely varied daytimes and nights are all caused through the manifestation of the primary divine Name '*the Light*' [III.201.35], which brings into manifest existence in the world all the transient images or 'likenesses' of the infinite divine Names:

> so when the divine Name '*the Light*' is considered (from the perspective of) the existence of the exalted shadow imaged forth (in the cosmos: i.e. the First Intellect or 'Perfect Human Being') and (from the perspective of) its rising upon those who are in the world, then the world (i.e. the creatures) which are in this image will call this rising, until the time it sets for them, a 'daytime', and from the time it sets for them, they call it a 'night'. But that (divine) Light is still present for this shadow, just as the Sun is still present in respect to the Earth both in its rising and in its setting (though it is visible to us only in its rising) . . . so in reality (that appearance of night) it is (only) a *shadow*, although they call it 'darkness'. . . . So know this!

> Then Allah made these days which we know, that are caused by the motion of the isotropic orb, and the daytime and night that are caused by the heart (of the cosmos) – I mean the Sun – in order to determine through them the effects of the *divine* Days that belong to the Names.
>
> [III.202.6]

Thus, as we have just seen, every divine Name has a specific Day (with its corresponding daytime and night), and, when the Day of one divine Name appears to be over, a Day of another Name starts, and so on; and all these are included in the eternal Day of the divine Name 'the Age':

> And every divine Name, known or unknown, has a (specific) Day in the Age, and these are the 'Days of Allah' – and all, in reality, are the Days of Allah, but most people do not know that. So if we descend from the divine Names to the Day of the First Intellect, we find that its effect in the Universal Soul divided it into daytime and night: its 'night' in respect to the Soul is when the Intellect turns away from her when he approaches his Lord to benefit (by receiving knowledge from Him), and its daytime in respect to this Soul is when he approaches her to benefit her, so this is her daytime. And through this effect Allah made in the Soul two forces: the intellective force, which is her 'night' in the world below her; and the active force, which is her daytime in the world below her – and that (contrast of these two universal powers) is called unseen and seen, letter and meaning, abstract and sensed. So this causes in the Soul a Day that has no daytime and night, while in the world it has (what we perceive as) daytime and night. The same applies to the Day of the 'Universal Matter' (*al-hayûlâ al-kull*): its daytime is its essence, and its night is its form, while it is in itself a Day with no daytime and night.
>
> [III.202.20]

This (i.e. the relation between the two forces of the Soul and the two phases of the day, the daytime and the night-time) leads to a very important conclusion that we (physically) *move* in the world in the 'daytime' (*nahâr*) of the Soul, and that we (psychically or spiritually) *perceive* the world in the 'night-time' of the Soul – a key symbolism of the Qur'an and Hadith that Ibn 'Arabî develops at length throughout the *Futûhât* and other works. This motion-perception that happens in the 'day' (*yawm*) of the Soul is a cyclic unit action that repeats every day (in fact, every *moment* for us). In other words: we either move or perceive, but not at the same time.

Likewise, the Day in every orb is a Day with no daytime and night for that orb, but with the appearance of a daytime and night for (some of) the world below it. Ibn 'Arabî even gives an excellent example, using a detailed scientific explanation about something that we are familiar with nowadays, the phenomena of daytime and night on the Earth:

> And Allah caused the (normal) daytime and night by creating the Sun and its (apparent) rising and setting on the Earth – whereas in the sky it is all light, with no daytime and night. The outlet of the night from the sphere of the Earth where the Sun sets is cone-shaped.
>
> [III.203.22]

This is exactly what everybody admits now as a fact of science, shown in Figure 2.2 for more clarification.

So with regard to us, we are necessarily either under the effect of the daytimes of particular divine Names of Allah or under the effect of their nights, depending on Allah's manifestation for each one of us and in everything else. But these Names in themselves have no daytime and night, but rather are all light. Ibn 'Arabî explains this differing influence of the different divine Names further by saying:

> So its night is the 'unseen' (*ghayb*), which is what is hidden away from us but at the same time affects the high Spirits which are above Nature and also the roaming Spirits; and its daytime is the 'seen', which is its effect in natural bodies down to the last elementary body.
>
> [III.201.15]

Ibn 'Arabî also gives more details about the different manifestations of God to different kinds of bodies and spirits, by dividing the daytime and night each into three thirds:

> and when Allah divided His Days like that, He made its night three parts and its daytime three parts; so He, the Exalted, descends down to His servants in the last third of the night of His Days [*Kanz*: 3355, 3388], and that is when He is manifested to the natural spirits (*al-arwâh al-tabî'iyya*) that manage the material bodies; and in the middle third He manifests to the subjected spirits (*al-arwâh al-musakhkhara*, or the angels of each heavenly sphere), and in the first third He manifests to the 'dominating spirits' (*al-arwâh al-muhaymina*).
>
> And He divided the day of these days into three parts, and in every part He is manifested to the world of bodies – for they are always praising Allah. So in the first third He is manifested to the subtle bodies (*al-ajsâm al-latîfa*)

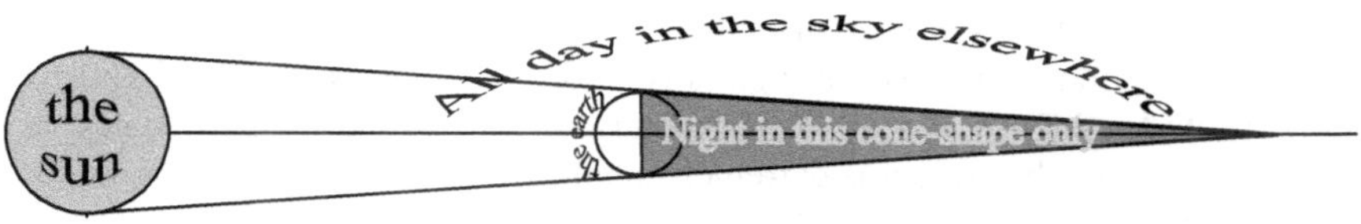

Figure 2.2 The daytime and night in the sky.

Note
Night is only in some regions below the corresponding orb, whereas in the orb itself it is all day. And in the case of the normal day on the Earth, the outlet of the night is extended in space as a cone.

which are unseen by sight, in the middle third He is manifested to the transparent materials (*al-ajsâm al-shaffâfa*), and in the last third He is manifested to the dense materials (*al-ajsâm al-kathîfa*). Without this manifestation, they would not be able to know Whom they are praising.

[III.201.24]

2.15 The 'single day'

As we have seen above, Ibn 'Arabî declares that there is an indivisible duration of time [IV.425.8] that is the smallest possible time or Day. Ibn 'Arabî calls this Day 'the single Day or time' (*al-zaman al-fard*) or 'the Day of task or event' (*yawm al-sha'n*). Another translation of *sha'n* is possibly 'concern', which could also convey the general meaning of the underlying Qur'anic verse, repeatedly cited by Ibn 'Arabî and it forms the basis of his unique view of the discrete nature of time: *kulla yawm Huwa fî sha'n* (55:29). But we prefer to use the words 'event' or 'task' in order to stress the meaning that in each such momentary 'Day', Allah acts in the world by creating the totality of all events. Ibn 'Arabî himself stressed this meaning: '(The *sha'n*) is nothing but the (single universal creative divine) Act, which is what He creates in each day of the smallest Days, which is the single time that is indivisible' [IV.425.11].

In every 'Day of event', Allah re-creates the world in a new image that is similar to the previous one, but with slight changes. Or in other words, in every 'Day of event' Allah causes one unique, singular Act in the world, because Allah is One and His Command (*amr*) is one (*Al-Masâ'il*, no. 26). However, this same single Act will have different results on the different entities in the world, depending on the capabilities and characteristics of each individual creature. For example, when Allah inspires the Universal Soul (*al-nafs al-kulliyya*) to move the element of fire (*'unsûr al-nâr*) in order to heat the world, the effects of this single Act depend greatly on the individual creatures, so those who are ready to burn will burn, and those who accept heat will be heated, and so on [*Ayyâm Al-Sha'n*: 11]: Also he says in the *Futûhât* that 'the task in relation to the Real is one from Him – but in relation to the recipients (*qawâbil*) of the whole world, it is many tasks (events) that – were it not (all) confined by existence – we could call infinite (because everything that physically exists is necessarily finite)' [II.82.6].

The 'Day of event' is a single and indivisible duration of time that equals the entire earthly global day at each instant, but because the Universal Intellect scans in this 'Day' N number of states (that is N/2 for bodies in its 'day' and N/2 for spirits in its 'night'), it appears to the observer, who is one of these states, as infinitely divisible, because it is extremely small: $24 \times 60 \times 60/N$ seconds, and N is surely unimaginably huge. This is simply because the observer exists only for this infinitesimal amount of time during every 'Day of event'. This infinitesimal amount of time is what we call '*now*' (*al-ân*), the '*moment*' (see the following section) or the '*presence*' – and this is the only *real* part of the imaginary time (see also section 6.8).

We must note that the isotropic orb, which is the first orb beneath the divine 'Pedestal', is not material. Therefore we can not differentiate between its 'parts', since there are no distinguishing features in this orb – which indeed is why it is called 'isotropic' (*atlas*), or the same in all directions. Therefore we can not measure the length of our normal day which is actually the 'day' of this isotropic orb (as we explained above in the previous section), and which also equals the Day of event. But rather we use this perceptible earthly day as a measure of other days, even though in itself it is an unknown duration of time. Although we divide it into 24 hours, each of which has 60 minutes of 60 seconds, all this is mere convention. We can not measure anything by its parts, so in order to measure this day, which is the period of the first orb, we need to compare it to other days 'before' it. But since it is the first day, we can not actually measure it: 'and when Allah created this first orb, it rotated a cycle which is unknown to other than Allah, the most Exalted, because there are no finite bodies above it, since it is the first transparent body' [I.122.29].

Moreover, Ibn 'Arabî also observes that since the isotropic orb has only one cycle, it can not be described as having an end, although we assume in it a beginning and an end [III.548.29].

In many cases Ibn 'Arabî calls this smallest, singular day 'the day of Breath' [II.520.33, III.127.335], and he even gives a measure of this day through the Breath: 'and the day is the magnitude of the breath of the Breather in the single time' [II.171.23]. This is because in this instantaneous 'Day' the divine creative 'Breath' emerges and returns to its Source, with its manifestation being the realm of all the divine 'letters', 'Words', or 'sounds' repeatedly produced by this Breath. As we shall see in section 7.8, Ibn 'Arabî visualizes the divine 'Word' as essentially composed of vibrations or sounds that are the 'letters' that form all the manifest objects and entities of the cosmos, just as letters form the words in human language. All words are composed of letters, and all letters are composed of the letter or initial out-breathing vowel sound *alif*, the first letter of the Arabic alphabet [I.78.22]. So in each single Day, this creative divine sound is produced by the divine Breath, and the manifest world is in reality the succession of these sounds or breaths. We shall explore this important issue in section 7.8 below. Also in section 6.8 we shall explain the creation scenario according the Single Monad model of the cosmos.

2.16 Moments

In addition to the common Arabic words used for 'time' – *zaman and zamân* – *al-waqt* is also widely used, which means the state 'you are always described by, so you are always under the rule of the moment' [II.538.32]. We have already mentioned above the difference between *zaman* and *zamân*, and shown that Ibn 'Arabî uses them basically in the same context, although he tends to use *zaman* for fixed and short times, such as the Single Time. *Al-waqt*, however, sometimes has a different meaning and technical usage for Ibn 'Arabî. He notes that:

> (this technical meaning of *al-waqt*) is in fact – by the convention of the Folk (the Sufis) – the state in which you are in the time being. So it is a thing that exists (now) between two non-existents (the past and the present; not as time but as things that have passed or are yet to come). And also it has been said that *al-waqt* is what comes upon them from the Real by (the Real's) managing (*tasrîf*) them, not by what they choose for themselves.
>
> [II.538.35]

Ibn 'Arabî elsewhere gives many equivalent meanings of *al-waqt*. For example he defines *al-waqt* in his short dictionary of Sufi terms (*Issttilâhât Al-Sûfiyya*: 8) by saying that: '*al-waqt* is your state in the time being, without any relation to the past or to the future'. But at the end he says that

> the divine (ontological) basis of *al-waqt* is His, the most Exalted, describing Himself as being *each day upon one task* (50:29): so *al-waqt* is what *He* is in – in the root – but it appears in the offspring that is the cosmos. So the 'tasks' (*shu'ûn*) of the Real appear in the entities of the world (the contingent things). So *al-waqt* in fact is what you are in, and what you are in is your aptitude itself. So what appears in you of the tasks of the Real that He is in, is only what your aptitude demands: so the task is already designated, because the aptitude of the possible with its possibility led the task of the Real to bring it into existence. Do not you see that the non-existent does not accept the existence (or the task of the Real), because it has no aptitude for that. So the (outward, manifest) origin of *al-waqt* is from the cosmos, not from the Real, and it is a kind of supposition (*taqdîr*) and supposition has no effect (*hukm*) on other than the creation (i.e. it has no rule on the Real).
>
> [II.539.2]

In this way *al-waqt* is the current moment of time; it is *our* portion of the single Day; so the single Day with regard to the whole world is the global reality encompassing all of manifest existence – including all on Earth – at any instant. But with regard to each entity in the world, it is a moment, since the time of each entity in the world is its *waqt*. This is why the Sufi was said to be 'the son of his moment (*ibn waqtihi*)', as indicated by Gerhard Böwering in his study of 'Ibn al-'Arabî's Concept of Time'. In this paper, Böwering also showed that 'In Ibn al-'Arabî's view there is an infinite cluster of moments, conceived as time atoms without duration, but they are mere instances of preparedness in which are actualized those possibilities that God has ordained to be affected in a human being' (Böwering 1992: 81).

However, the issue of whether the moment has a duration or not is extremely delicate. If we choose to say it has not, then how can the extent of the entire perceived day be composed of zero-length moments? We must keep in mind that the number of entities in the world (N), though very big, is finite. Therefore the single Day equals N multiplied by the duration of the moment. But if we choose to assert the actual duration of the 'moment', it would lead to questions already encountered in earlier philosophers' paradoxes about time, such as 'What exactly happens *during* the moment?' and 'Does the event that takes a moment *pop up* at once or gradually?' For if at once, then 'Why does it take a duration?' and if gradually, then 'What are the sub-events, and how many are they?'

Ibn 'Arabî did not give any direct insight about this subtle and highly important cosmological issue. However, in his cosmological treatise *'Uqlat Al-Mustawfiz*, he spoke briefly about the Greatest Element (*al-'unsûr al-a'zam*) whom Allah created at once, and we shall show in section 6.4 that the Single Monad, though it is an indivisible unit, is composed of or made by (or from different manifestations of) this Greatest Element. Therefore, the moment (that corresponds in fact to the creation of the Single Monad) should also be composed of 'sub-moments' (that correspond to the creation of the Greatest Element). We can now affirm, according to Ibn 'Arabî, that those sub-moments are utterly indivisible because he said that Allah creates the Greatest Element 'at once'. The questions of 'Do those sub-moments have non-zero durations?' and 'How many sub-moments are in the moment?' remain open, though a first speculation is that the process is similar to the normal day where the Sun rises and sets to define the daytime and the night. There is some support for this in the comparison Ibn 'Arabî usually makes between the creation of the Perfect Human Being (that is the Single Monad) and the creation of the world, and more specifically the motion of the Sun during the day. We shall talk about this comparison in section 6.7. Therefore, the Day of the Single Monad (that is the moment) may look smooth and composed of a continuous flow of time. But it is quite possible that the same sets of questions may be repeated and similar quantization takes place at smaller scales.

When the Single Monad (that is the Universal Intellect) faces his Lord, this is the 'night' for the Universal Soul; and when the Intellect (*'aql*, masculine in Arabic) faces the Universal Soul (*nafs*, a feminine noun) this is her daytime [III.202.22], so this is the single Day of the Single Monad that is the Universal Intellect, and therefore the moments that are the days of the sub-entities of the world (which are the sub-intellects) should be also in the same way. But we have not seen any reference in Ibn 'Arabî's writings with any detail about the exact relation between the Single Monad and the Greatest Element, and hence about the moment and its possible constituents. On the contrary, Ibn 'Arabî affirms that this is a divine secret and that he was sworn not to disclose it (*'Uqlat Al-Mustawfiz*: 38).

2.17 The future, the present and the past

We have already said that the future and the past do not exist; the real time is only the present, the now (*al-'ân*) or the current state (*al-hâl*). Also, we explained that the Age is an endless circle that does not have a beginning or an end for itself; it is all an 'eternally ongoing present' which is called *al-ân al-dâ'im* or – as Ibn 'Arabî calls it elsewhere – 'the continuous existence' (*al-wujûd al-mustamirr*) [II.69.13, IV.362.32]. However, we live in time at the present now, and we feel the past and the future, so how can they be non-existence?

As we noted earlier, Ibn 'Arabî often explains that the essential realities [*a'yân*: the essences or entities] of the manifest world have always been existing in the Knowledge of Allah, though not actual real existence [II.309.25, I.538.32]. When Allah creates the world, it appears in real existence in the same reality as it is determined in Allah's knowledge. Allah does not create all the successive states of the world at once, but in a series process; so the creations are brought into existence one by one. Of course this is true in relation to us, but with relation to Allah, Who is out of time, the word 'series' would be meaningless because it is confined to time. For Allah nothing was changed because the creation existed in His knowledge eternally. So because we are sub-entities in this whole creation, we encounter time as past, present and future, but in fact only the present exists. The past is 'a relegation of an actualized non-existence', and the future is 'a pure non-existence' [II.69.13], whereas the present now (*al-ân*) or state (*hâl*) is 'what makes the distinction between them' [II.56.11], 'so without the (current) state (*hâl*) there would be no distinction between the past non-existence and the future non-existence, so the now (*al-ân*) is like a partition (*barzakh*)' [III.108.16]. 'Therefore the present state (*al-hâl*) is described by the continuous existence and it is the constant and immutable rule, and anything other than the present state is non-existence and could not be a firm (absolute) existence' [IV.362.33].

2.18 Eternity

As we have already noted, the English word 'eternity' has two different Arabic synonyms: '*azal*' and '*abad*'. *Azal* is eternity without beginning (*a parte ante*), and *abad* is eternity without end (*a parte post*). Ibn 'Arabî showed in his short treatise The *Book of Eternity* (*Kitâb Al-Azal*) that 'there is nothing called eternity at all' (*Kitâb Al-Azal*: 8–9). Eternity is in fact the *negation* of a beginning (or end), not endless extension: that is why Ibn 'Arabî says that eternity is a negative attribute. In this way there is no meaning to asking whether there has been any extension of time between the existence of Allah and the existence of the world, because Allah creates the world out of time, as discussed in section 2.3 above.

There has been a long debate amongst Islamic philosophers and theologians about the meaning of eternity and time with relation to Allah. Some philosophers say, for example, that Allah was talking in eternity in His eternal Speech, and that therefore He said in eternity '*take off thy shoes*' (20:12) to Moses, and '*worship thy Lord until certainty comes unto thee*' (15:99) to Muhammad, peace be upon them, and so on. They say that because clearly we can not say that Allah uttered these words at the time of Moses or Muhammad, therefore it must be an eternal speech. Ibn 'Arabî, however, showed in his *Kitâb Al-Azal* that all these interpretations are not proper, and that they end up confining Allah to time, which is a serious error: 'It is more proper to say that Moses heard out of time, because the Speaker spoke out of time. To let Moses become holy is better than letting *al-Bâri'* (the Creator) be compared to us' (*Kitâb Al-Azal*: 5, see also *al-Masâ'il*, no. 131).

So if we take the correct meaning of the word 'eternity', which is the negation of any beginning (*azal*), we may ask: was there anyone in eternity with the Creator or not? There has also been a long philosophical and theological debate about that, which Ibn 'Arabî summarizes by saying:

> One group said: 'the primordials (*qudamâ*) are four: the Creator, the Intellect, the Soul and the Dust (*al-haba'*)'.
>
> Another group said: 'the primordials are eight (pre-existents): the Essence and the seven descriptions' (i.e. the seven primordial descriptions or attributes of Allah, drawn from earlier kalam theology: 'Life, Knowledge, Ability, Will, Hearing, Seeing and Speaking' [I.525.32]).
>
> Another group said: 'nothing is primordial but One, and He is the exalted Real, and He is One in all aspects, but to His Essence (there is) an aspect by which He is called Able, and so on for whatever they have made a description'.
>
> Another group took this (last) opinion, but they added to it a (new) concept. And this concept is called 'the Reality of Realities', which is neither existing nor non-existing, but it is primordial with the primordial and created with the created; it can be imagined, but it does not exist by itself, like universality (*'âlamiyya*) and so on.
>
> (*Kitâb Al-Azal*: 8–9)

Clearly, the Sufis are among the last group, and Ibn 'Arabî in particular relates the beginning of creation to 'the Reality of Realities' (*haqîqat al-haqâ'iq*), and sometimes he calls it 'the Universal Reality' (*al-haqîqa al-kulliyya*) or 'the Muhammadan Reality' (*al-haqîqa al-muhammadiyya*). He stated in chapter 6 of the *Futûhât* that the beginning of the spiritual creation is the 'Dust', and that the first existent within it was the 'Muhammadan Reality of (divine) Mercy' (*al-haqîqa al-muhammadiyya al-rahmâniyya*) that is not confined to space, and that it is created from the 'Known Reality' (*al-haqîqat al-ma'lûma*) that can not be described by either existence or non-existence [I.118.5]. Ibn 'Arabî claims that only the Sufis have introduced the concept of 'the Reality of Realities' (*haqîqat al-haqâ'iq*), although he admits that the Mu'tazilites drew attention to something similar to this notion when they tried to escape the accusation that their understanding of the divine Attributes postulated the real existence of additional realities other than the Essence of the Real [II.433.14, *SPK*: 134–139].

2.19 The age

So the days are repeated cycles of different 'orbs', whether celestial or divine: every orb has its own day. These days overlap with each other, so the day of the Sun, for example, takes 360 days of our days that we count on the Earth; the day of the Moon takes 28 normal Earth days, and so on. In one Sun day, many Moon days pass; and in one Moon day, many normal Earth days pass – just as in one 'Lord-day', one thousand Sun days (Earth years) pass. Likewise, *all* the possible days of *all* orbs and divine Names happen and repeat themselves in the 'Day of the Age' that 'is one Day that never repeats, and it has no daytime (*nahâr*) or night' [III.202.5]. The Day of the Age is *all* a 'daytime': it does not have a night, because there is no orb above it; it is the truly all-encompassing orb that includes all material and spiritual orbs.

Ibn 'Arabî, like other Sufis, consider that 'the Age' (*al-dahr*) is one of the divine Names of Allah, according to a hadith in which the Prophet Muhammad says: 'Do not damn the Age, for Allah is the Age' [*Kanz*: 8137]. In another divine saying (*Hadith qudsî*), Allah says: 'the son of Adam hurts Me: he damns the Age, and I *am* the Age, the command in My Hand, I circulate the daytime and the night' [*Kanz*: 8139].

Also in the Qur'an, Allah says: '*and they say: What is there but our life in this world? We shall die and we live, and nothing but the Age can destroy us. But of that they have no knowledge: they merely conjecture*' (45:24).

In his comments on this last verse, Ibn 'Arabî affirms that they (the disbelievers) are quite right in their claim that nothing but the Age can destroy them, because Allah *is* the Age – though they did not mean that, but rather meant simply time, and not 'the Age' as a name of Allah [IV.265.29].

We have seen above that 'the time of a thing is its presence': this is why Allah is named as 'the Age', because the realities of everything exist in Him. So He is 'the Age' because the whole world (hidden and manifest), including all space and time, from eternity without beginning (*al-azal*) to eternity without end (*al-abad*): all this is a manifestation of Allah's divine Names – so the Age is the all-encompassing orb of all Names [IV.266.11]. Ibn 'Arabî explains this by saying:

> Do not you see in Allah's words (the Qur'an) when He told us about things in the past, He then used the past tense, and He used the future tense for things to come, and He used the present tense for the now-happening things, . . . and if we seek for all that something that has a real existence and these things are occurring in it – which is for them like a container – we shall find nothing, neither by mind nor by sense, except that we find it as an imaginary container, which is itself contained by another imaginary container, and so on, nothing but endless illusion. So if you think rationally, you will find that there is nothing (i.e. no 'containing' reality of space or time) that can be comprehended by imagination or by the mind, nor by the senses, nothing but the Real Existence on which our existence is based.
>
> For this reason He named Himself to us as 'the Age', so that the reference can be only to Him, not to 'time' as imagined, because there is no Ruler (*hâkim*) other than Allah, and in Him appeared the realities of things and their properties. So He is the everlasting existence, and the realities of things with their properties appeared from behind the veil of His existence, but because of His subtleness (*latâfa*), we see the realities of things, which are ourselves, from behind the veil of His existence without seeing Him, just like we see the stars from behind the veil of the heavens (the orbs) and yet we do not see the heavens.
>
> [III.547.1]

Moreover, given the above definition of the Age, it is not possible to imagine a plural form of this word in this meaning. Yet Ibn 'Arabî himself sometimes uses it as plural, when he speaks about the 'first age' or the 'age of ages' from which time appeared: 'from that (eternity) is the Prophet's saying: "Allah is/was, and none with Him" (*Kanz*: 29850), and this is a rare knowing known only by the "Solitary Sages" (*al-afrâd*) among men. That is what is called "the first age" and "the age of ages": and from this eternity (*azal*), time emerged' [I.156.35].

This could be slightly confusing, but there are some references in Ibn 'Arabî's writing that seem to parallel what is known in modern cosmology as the 'bubbles' model of the universe, where the universe is full of black holes that cause a large curvature in space-time in such a way that each can be considered as a separate space-time similar to – and as large as – our own; except that each is only a single point in our space-time. Thus Ibn 'Arabî affirms that 'each part of the world can be a cause for another world similar to it' [I.259.25]. Also Ibn 'Arabî considers that the origin of the world that we live in is the Universal Intellect, who is originally only one of an unknown number of roaming Spirits whom Allah created directly without any intermediaries [III.430.5]. Any one of these primordial Spirits is, therefore, capable of making another universe, just like the Universal Intellect that made our universe. Also we may recall that Allah in the Qur'an is repeatedly called the 'the Lord of the world*s*', in the indefinite *plural*, and not only of this world. However, the Arabic word for 'the age' has also been conventionally used sometimes for other related meanings such as eternity without beginning (*azal*) and eternity without end (*abad*); so Ibn 'Arabî showed that the plural form 'ages' (*duhûr*) could also be used to include or refer to all these terms, since they all are essentially the Age [IV.266.5].

On the other hand, the Age is not only time, but it is also space. Because the Age is the totally 'encompassing orb', it must include everything inside it, spatial as well as temporal. As we shall show in section 3.6, the world as both space and time is created in the seven days of the divine Week (six Days for space, plus Saturday for time). And the Age is not more than those divine seven Days [II.438.5]. Thus in Ibn 'Arabî's conception, space and time have the same meaning as days, and both are unified as space-time, where the divine Week is its basic unit, and the Age is in fact this Single Week.

2.20 Other expressions of time

In addition to the common words such as time (*zaman, zamân*), moment (*waqt*), day (*yawm*), daytime (*nahâr*) night (*layl*), eternity (*azal/abad*) and the age (*dahr*), which we have discussed above, along with the related terms for week (*usbû'*), month (*shahr*) and year (*sana*), which we shall discuss in Chapter 3, there are other time expressions occasionally used by Ibn 'Arabî with slightly different and more specific technical meanings than in their usage by other Muslim scholars and theologians. We want to end this chapter by looking at some of these other temporal terms for the purpose of completeness, although we may not need to refer to these technical terms in the rest of this book. In particular, some of these expressions became important in later forms of Islamic thought which tried to elaborate and integrate Ibn 'Arabî's ideas, in various ways, with the conceptual schemas of both kalam theology and Avicennan philosophy.

- *Al-sarmad*: this is another word for eternity, other than *al-azal*, *al-abad* and *al-dahr*. Unlike kalam theologians, Ibn 'Arabî does not use this word often. *Sarmad* means 'absolute eternity', i.e. it includes both eternity without beginning (*azal*) and without end (*abad*). It is most widely used as an adjective, *sarmadî*, meaning 'everlasting', and this is how it is used in a very pertinent discussion of time in the Qur'an (28:71–72). Ibn 'Arabî normally used it [I.164.2, I.169.26, II.675.14, IV.29.27] in a similar manner.
- *Matâ*: literally this word means 'when?', and refers to the relation of a thing or event to time, or to other events. It is used to inquire about the time or occurrence of an event in relation to others. Ibn 'Arabî affirms that the thought of time comes about due to this kind of inquiry: 'So for example if one asks: when (*matâ*) did Zayd come? Then the answer may be "(he came) when (*hîna*) the Sun rose"' [III.546.28].
- *Al-hîn*: this is a (relatively short) duration of time [II.201.36, II.263.25]; it is also commonly used to refer to a specific future or past time [II.201.36], usually in answer to 'when', as in the previous example: '(he came) when (*hîna*) the Sun rose' [III.546.28].

3 The significance of the divine week and its seven days

The elements are four mothers,
and they are the daughters of the world of (the heavenly) orbs.
We are born out of them, so our (ultimate, spiritual) *being*
is in the world of (spiritual) principles (arkân) and (spiritual) rulers.
The God has made our sustenance from the ears of grain (sanâbil),[1]
from the influence of Sunbula (Virgo),[2] *without sharing.*[3]
And He also multiplied our reward by 'the seven ears of grain',
as it was said in a saying (2:261) *by One Who does not lie.*[4]
So our time (zamân) is seven thousands (of years),
that come through the recurrence of (moments) of radiance and intense darkness.
So see, with your intellect, seven in seven,
from seven that are not kings.

[I.292.31–293.3]

As Ibn 'Arabî suggests in this cryptic opening poem of his key cosmological chapter 60 of the *Futûhât* (explained at the end of section 3.1), although we encounter many visible, apparent 'days' due to the rotation of the Earth around its axis, which all appear to be similar to each other, ultimately we can reduce them to only seven distinctive Days[5] depending on the kinds of events that happen in them. Ibn 'Arabî argues that in each Day of the seven Days of the (actual, *cosmic*) 'Week', Allah orders the heavenly orbs to act in a special unique manner that causes unique events and motions to appear in the entire cosmos. However, the seven Days of these divine, creative 'events' or 'tasks' (*shu'ûn*) are intertwined with our normal days of the apparent earthly week in a special manner that we shall explain in the coming chapter – and that is why we see multiple events appearing every day.

In various sections of the *Futûhât*, Ibn 'Arabî goes on to explain the importance of the divine 'Week' (*usbû'*) and the different meanings of the seven Days of that Week, which are rooted in the meanings of the Arabic terms for each of those Days, running from 'the First' (*al-ahad*, corresponding to Sunday) to 'the Day of rest' (*al-sabt*, corresponding to Saturday). He also shows that Saturday is therefore the '*Day of eternity*' (*yawm al-abad*) and that all the Days of the Week of creation, including Saturday itself, are actually occurring on this 'last' Day of eternity (*al-sabt*). These distinctive conceptions can only be fully understood after investigating Ibn 'Arabî's view of the actual flow of time, which we shall explain in the following chapter; but in this chapter we shall first explain the origin and the meaning of these seven Days of the divine creative Week, why they have to be seven and not more or fewer, and what is the significance of each Day.

Also in this chapter we shall explain Ibn 'Arabî's unique understanding of the process of creation of the world by Allah '*in six Days*', from Sunday to Friday. He explains that the creation of the world in this Week corresponds to the creation of the 'six directions' of space (in six Days, from Sunday to Friday) and then the appearance of this creation *in time* on Saturday (the Day of eternity). In this way the Week is indeed the unit of the space-time container of the world – or also 'the Age' as we explained in section 2.19. This distinctive conception of the unification of space and time, symbolically expressed in Ibn 'Arabî's understanding of the seven Days of creation, adds an essential dimension to Relativity's concept of space and time; as a result, for the first time in astronomy, the 'Week' will itself have a clear and essential cosmological significance.

3.1 The significance of the week in theology and astronomy

In most Islamic books that talk about cosmology, we find many diagrams and tables that associate each day of the seven days of the week with specific letters of the alphabet and specific divine Names of Allah, in addition to certain planets and constellations or zodiacal signs. For example, Ibn 'Arabî explained in the long chapter 198 of the *Futûhât* [II.390–478] the creation of the world by Allah and the role of His divine Names on the different parts of Heavens and Earth, and then he relates each divine Name and the thing He creates to a letter from the Arabic alphabet, a mansion from the 28 lunar mansions (constellations), a day of the seven days of the week, and one of the seven circulating heavenly bodies (five planets, Sun and Moon). This type of symbolic association based on the days of the astronomical week is also found in many other cosmological books of other religions and cultures dealing with astrology and related mythology.

The Egyptians once divided their 30-day months into three ten-day weeks, in the same manner as Greeks of the same period (Goudsmit 1966: 24), but later they changed it back to seven days.[6] More recently, the French (during the Revolution, in 1792) tried to make their week ten days instead of seven; and the Russians also tried (in 1929) five-day and (in 1932) six-day weekly systems, although they all later restored the seven-day week (Goudsmit and Claiborne 1974: 24).

So what is the importance of the week, and why does it apparently have to be seven days in particular, while most other attempted systems making the week three, five, six or ten days did not persist, with few exceptions such as the Maya who used weeks/months of 20 days (Aveni 1990: 101, 185–252)? Even some micro-organisms clearly adopt a seven-day biological cycle (Aveni 1990: 100, and Coveny and Highfield 1990: 220–259). And yet, unlike the day, the year and the month, there is no any apparent *astronomical* significance to the week; nothing cosmic happens in the heavens in seven days.

The Babylonians and the ancient Egyptians believed that each hour of the day was ruled by one of the five then-known planets plus the Sun and the Moon, and they named the days of the week after the names of these seven circulating planets, as they are now used in both the Latin and Anglo-Saxon languages and cultures (Aveni 1990: 102–106). They considered that the planet that ruled the first hour of the day governed the entire day, so they gave the name of this planet to the corresponding day. The same doctrine was also part of earlier Persian cosmology and theology (Bickerman 1968: 59). Ibn 'Arabî also refers to the same hypothesis [III.203.31].

It is interesting to note that these celestial bodies, in the same sequence, were also used to name the days of the week in ancient India, Tibet and Burma (Parise 1982: 172). This is also true for the names of Japanese (who used the Chinese sexagenary cycles) (Parise 1982: 215–218) days of the week, but the custom there has been traced back only 1,000 years. Adherents of the cult of Sin at Harran, who were known as Harranians or 'Sabeans' by Arabic and Syrian authors, also named their days after the same solar system members (Langdon 1964: 154). Like Ibn 'Arabî, the Babylonians, the Chinese, ancient Egyptians and most ancient civilizations considered the day named after Saturn to be the seventh day, so they began their week with a day named after the Sun (Sunday), a practice which was affirmed in the Bible[7] and later by the Prophet Muhammad.[8]

In the Arabic linguistic usages followed in Islam, however, the names of the days of the week do not relate to the names of any pagan gods or celestial bodies. Before Islam, different names were used in Arabic, which were mostly derived from certain actions people usually performed on those particular days of the week, though some of those Arabic names might also have been derived from the names of the planets (Al-Marzûqî 2002: 238–244). But in the later standardized Islamic usage, apart from the day-names *Jum'a* (Friday) which means 'gathering' and *Sabt* (Saturday) which means 'rest', the names of the days are merely numbered from one (*al-ahad* which means 'the first' or 'the one', for Sunday) to five (*al-khamîs* which means 'the fifth' for Thursday). These names, however, clearly suggest that Sunday (*al-ahad*) is the first day of the week, as was the case with the earlier Babylonians and Egyptians – a fact which Ibn 'Arabî and some other Muslim authors normally take for granted, based on many related prophetic narrations, as we shall see further below (section 3.5).

Given their centrality in Qur'anic accounts of the creation, the seven days of the week (and their standard Arabic names) play an essential role in Ibn 'Arabî's cosmology. His cosmological understanding of the week clearly has its basis in the scriptural accounts of divine creation, but also assumes throughout that there must be some kind of corresponding deeper effects of that (i.e. the divine creation of the seven Days of the Week) in the wider cosmos.

Finally, although there are many similarities between Ibn 'Arabî's doctrine about the origin of the divine creative Week and its Days, and the cosmological perspectives and understandings of earlier ancient (mostly pagan) cultures, he also takes great pains to stress that this cosmological schema should not be understood as a deviation from the fundamental monotheistic teachings of Islam, and to indicate the ways that conception is rooted in indications in both the Qur'an and many hadith. Although Ibn 'Arabî's cosmology, for example, relates the seven days of the week and the orbs of the seven moving planets, as in many ancient cosmologies, he carefully emphasizes that he does not consider these planets as 'gods' at all. Thus he goes on to explain, in his explanation of the poem (quoted at the beginning of this chapter) opening chapter 60 of the *Futûhât*, that the planets and/or the angelic spirits associated with them 'are servants, and the servant does not deserve the name "ruler" (or "king": *malik*). And the "seven" mentioned (there) are the seven planets in the seven orbs that appeared by the seven Days of the Week' [I.293.4].

Following repeated indications in the Qur'an and Hadith, Ibn 'Arabî understands that those planets, along with other constellations associated with signs of the zodiac and lunar mansions, are associated with or inhabited by certain spirits (*rûhâniyyât*) or angels whom Allah appointed and organized in a specific hierarchy to look after the whole cosmos beneath them, including the Earth [III.433–434]. This is different from earlier cosmological doctrines, because the pagan astrologers believed that these spirits were deities and gods, while Ibn 'Arabî stresses that they are nothing but servants created and appointed by Allah.[9]

3.2 The four main time cycles

Ibn 'Arabî stresses that 'everything in the world has to be based on (specific) divine Attributes' [I.293.5]. Although some Muslim scholars, following a famous hadith [*Kanz*: 1933, 1937], believe that the basic divine Names or Attributes of Allah can be limited to 99, Ibn 'Arabî considers them to be countless [III.146.35], while the 99 Names that are referred to in some prophetic narrations are simply the main most Beautiful Names (*al-asmâ' al-husnâ*) of Allah. Of these many divine Names, there are four fundamental Attributes – Life (*hayât*), Knowledge (*'ilm*), Ability (*qudra*) and Will (*irâda*) – that are necessary and sufficient for Allah to be described as God. Therefore those are considered to be the ultimate sources or 'mothers' (*ummahât*) of all other divine Attributes [I.469.25]. In relation to creation, however, three more Attributes are also necessary for Allah to be Creator: Hearing (*sam'*), Seeing (*basar*) and Speaking (*kalâm*). Together, that makes the principal divine Attributes of Allah to be 'seven mother attributes. . . .: Life, Knowledge, Ability, Will, Hearing, Seeing and Speaking' [I.525.32].

Because Allah created (the perfect) Human Being 'according to His Image' [I.163.20], these same divine Attributes are potentially manifest in every fully human person (such as Adam and the prophets). Also, as Ibn 'Arabî says, Allah created the world and everything in it in the image of (the Perfect) Human Being [II.652.25], and so the world with the Human Being is 'on the Image of the Real'

– but without the Human Being it would not have this perfection [III.343.25]. So these same attributes should be available and essential in the world as well. That is why, he explains, the numbers four and seven play a central role in the world: the four elements in nature (earth, water, air and fire, already mentioned in his poem opening this chapter), the four time cycles, the seven heavens, the seven days, and so on. The two cosmologically fundamental four fold groups that emerged out of the four 'mother' Attributes (Life, Ability, Will and Power) that are the four aspects of the divine Presence of the Essence (*al-Dhât*) are the four earthly elements (earth, water, air and fire) and the primordial cosmological principles of the Intellect, Soul, Dust and Nature, as in Figure 3.1.

This quadratic cosmological rule was also reflected in relation to time. Therefore, Ibn 'Arabî points out, there are four main time cycles within the domain of manifest nature: the day, the week, the month and the year. These four *natural* time cycles have their origin in the effects of those four elements of Nature (fire, air, water, earth) that are originally derived from the above-mentioned four principal divine Names ('the mothers'). As Ibn 'Arabî says:

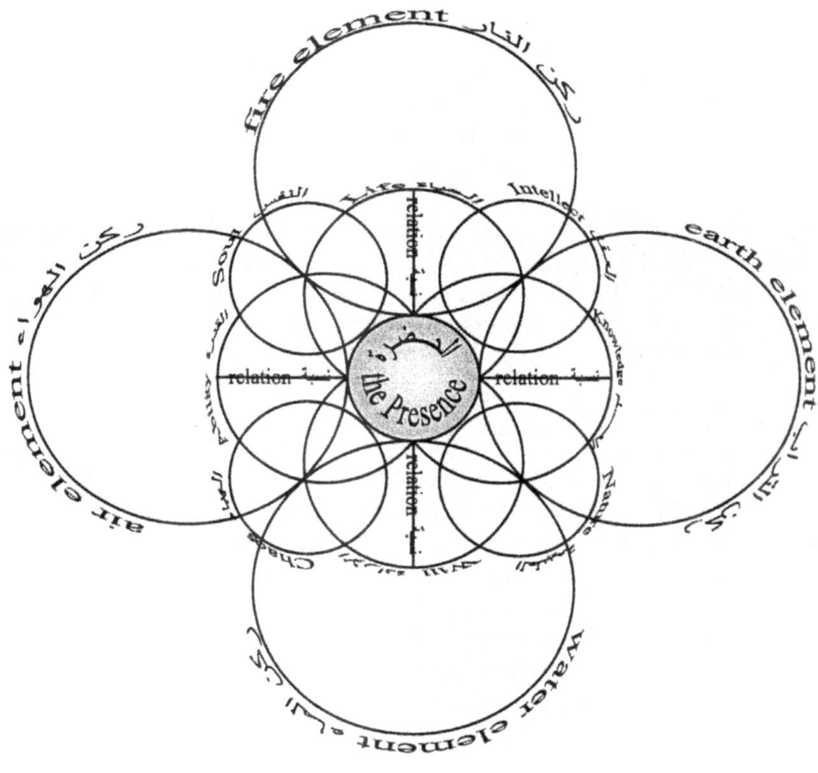

Figure 3.1 The Divine Quadratic Rule.

Note
This figure is translated from the *Futûhât* [I.260].

time is restricted to the year, month, week and day. Time is divided into four divisions because the natural seasons are four, because the origin of the existence of time is Nature, whose level is below the (universal) Soul and above the 'Dust' (*habâ'*) that philosophers call the Universal Matter (*hayûlâ*). The influence of this (principal) quaternity (*tarbî'*) in Nature is from the influence of the (same principle of) quaternity in the divine influences from (the fundamental Names) Life, Knowledge, Ability and Will. For by these four (Names), godship is confirmed for the God.

So the quaternity (first) became manifest in Nature. Then the (divine) Command descended until the (principle of quaternity) appeared in the 'biggest time' (cycle), which is the year, so that it was divided into the four seasons: spring, summer, autumn and winter. This was brought about by the motion of the Sun through *the stations* (of the zodiac), which have been divided by Nature into their (seasonal) divisions according to the (natural) elements that are the 'basic principles' (of fire, air, water and earth).

[chapter 390, III.548.17]

3.2.1 The day

We have already discussed Ibn 'Arabî's concept of the 'day' in Chapter 2, where we showed that he defines the 'day' as all that is included within the revolution of the Isotropic Orb, which encompasses all of material existence. This day is astronomically defined by the rotation of the Earth and it is conventionally divided into smaller units such as hours, minutes and seconds (see also section 4.6). The divine Day, however, is the corresponding effects (manifestations) of each of the seven fundamental divine Names on the entire cosmos, as we shall see further below (section 3.4). This unique divine Day is in fact the smallest indivisible unit of time, though it equals in length the normal day as we discussed in section 2.15.

3.2.2 The week

The second time cycle is the week, which Ibn 'Arabî – following the detailed Qur'anic indications – considers to be the main cycle of Creation. The week (which is seven days) has its origin in the seven main Attributes of Allah, but until now it does not seem to have any particular astronomical significance. However, Ibn 'Arabî's unique view, as explained below, gives a profound and essential significance to the Week in terms of astronomy/cosmology as well as the theology of creation. This will be the main focus throughout this chapter (see in particular section 3.3 below).

3.2.3 The month

Ibn 'Arabî distinguishes between the witnessed lunar month, which is from new moon to another new moon, and the 'divine Month' which is the time needed for the Moon to perform one full revolution in the orb of the zodiac: that is, as Ibn 'Arabî says, 28 days [III.548.28]. He also recognizes the solar 'month' as the Sun's observed motion throughout the zodiac, where the zodiac is conventionally divided into 12 parts, each corresponding to one month [I.388.20], though he does not give any details about the length of solar months in terms of their days.[10]

3.2.4 The year

The year, for Ibn 'Arabî, is the time needed for the Sun to perform one full revolution in the orb of the zodiac [III.548.28], as witnessed from the Earth. Like the Babylonians,[11] Ibn 'Arabî considers the year to be 360 days [III.434.9], and not like our calendar year of 365.25 days.

Ibn 'Arabî regards our solar year and the solar (and lunar) month as conventions set up by human observers, while the 360-day year, the 28-day month, the (seven-day) week/Week, and the (sidereal) day/Day are divine periods of time set up by Allah when He created the heavenly orbs and made them move [III.548.27]. It is noteworthy in this regard that the 360-day year does not equal 12 of the 28-day months. These four time cycles that Ibn 'Arabî talks about are not meant for calendar purposes; they are said to be the actual measures of time set up by Allah when He created the world. Moreover, Ibn 'Arabî shows that this non-integer ratio is preordained and essential for the vastness of creation, because the creation is built upon the act of generation (*takwîn*), and with complete ratios no generation could happen; so there have to be integers and fractions [II.440.7].

The differences between the witnessed lunar month (synodic lunar month = 29.53 days) and the divine lunar month (of 28 days), and between the witnessed year (365.25 days) and the 360-day year, might be because of the interference of the different motions of the Sun, the Earth and the Moon. As the Earth spins around its axis it also rotates around the Sun, and as the Moon rotates around the Earth it also moves with the Earth around the Sun. These interfering motions may account for the difference. For example, if we measure the period of the Moon relative to those stars that are apparently fixed (this is called the sidereal lunar month), we get only 27.32 days (and not the usual 29.53 days, the observed lunar month). The divine lunar month for Ibn 'Arabî is 28 days because he measures that period in relation to the zodiac (far-away galaxies) or actually the Isotropic Orb, and not the orb of the Sun or the constellations [I.656.13], because those constellations are not actually fixed [III.549.3]. Also we have to know that the length of the observed earthly day varies from one place to another on the Earth and from summer to winter throughout the year, and that the normal solar year and the normal lunar month slightly vary from time to time owing to the influence of gravitation of other planets and stars that change their positions inside and relative to the solar system, respectively. Thus the mean solar day in the year 2000 is about 1.7 milliseconds longer than it was in 1900, and is slowly getting longer. There is a possible allusion in the Qur'an to such long-term changes in the length of the year and the month, where Allah says: '*The quantity of months with Allah is 12 months* (in a year) *by Allah's ordinance in the day that* (when) *He created the Heavens and the Earth*' (9:36): in other words, it could be that the year started as 12 months each of 28 days, but that this then changed with time as the motion of the Earth slowed down and as the solar year became longer, and also as the lunar month became longer than 28 days.

In this regard, we should also notice that in Arabic there are two names for the year which do not appear to have identical meanings: *sana* and *'âm*. Although both terms are currently used to refer to the year, it seems from the etymological meaning of those two names and from Ibn 'Arabî's and Qur'anic usage that the word *sana* means the original 360-day year, while *'am* – which literally means 'entire' or 'full' – is the time needed for the Earth to make a full revolution around the Sun, which is the slowly lengthening conventional year now observed on Earth. In the Qur'an, Allah distinguishes between these two Arabic words in one verse that declares the time that Prophet Noah stayed with his folk: '*And verily We sent Noah unto his folk, and he continued with them for a thousand years (sana) save fifty years ('âm); and the flood engulfed them, for they were wrong doers*' (29:14).

3.3 The week as the primary time cycle

While Ibn 'Arabî considers the Week (of Creation) to be the primary time cycle, only the week among these four cycles does not seem to have any apparent astronomical significance. We can say only that the week is one-quarter of the divine lunar month ($28 = 4 \times 7$). From the observed astronomical point of view, the day should be the primary time cycle, because it is the smallest standard period of time as far as the solar system and the Earth are concerned, and all other three cycles (as defined by Ibn 'Arabî) are integer multiples of the day, while the year is not an integer multiple of the week. However, we shall see that Ibn 'Arabî does not consider the day to be the primary cycle because the Days of the divine Week are not similar to each other, as they might appear to us. Since each Day of the Week is based on one of the seven fundamental divine Attributes of Allah, so these Days are not identical because those seven divine Attributes are not identical. Therefore the Week, rather than the day, is the primary cycle of divine time, and each day of the seven Days of that Week is ruled by one of the seven fundamental divine Attributes.

However, in keeping with Ibn 'Arabî's essential understanding of the 'ever-new creation', this does not mean that any particular day of this week is *identical* to that of another week. They are only 'similar' to each other because they are originated from the same divine Attribute. Ibn 'Arabî says:

> Nothing is actually repeated, because of divine vastness (*ittisâ'*); so (everything) is in ever-new, not renewed, existence. Thus if we call the new (thing) 'renewed', that is because it is extremely similar (but not identical) to its counterpart, so that they can not be distinguished from each other . . . and

> the daytime and night are called 'the two-new' (*al-jadîdân*), and not 'the two-renewed' (*al-mutajaddidân*), because Saturday is not Sunday and it is not Saturday from the other week, or from another month or from another year.
>
> [III.127.23]

This is clearly evident in modern astronomy, because whatever periodical motions we see locally in our solar system are actually part of a more global motion that, in the end, never repeats itself in the same way, because everything is moving (see sections 1.1 and 1.4). In fact, Ibn 'Arabî always stresses that there can not be any two identical forms in the world, and that this is because 'Allah never manifests in the same form twice, nor in the same form to any two persons' [III.127.33].

Therefore Ibn 'Arabî maintains that 'although there are many days, the real order of events reduces them into seven days' [*Ayyâm Al-Sha'n*: 6], which are the seven days of the week; and then these days iterate in months and years. And as we showed, this is due to the fact that '(the main) divine Attributes are seven, not more, which made the Age not more than seven (distinctive) Days' [II.437.30].

However, the observed, earthly week and its days that we witness and live through do not seem to be distinctive in any natural way; as noted earlier, they appears to be purely conventional. The reason for that is the 'intertwining' between the underlying divine seven 'Days' of creation and the days that we live. This intertwining of the two kinds of days is a complicated concept that Ibn 'Arabî explained partially in his short book *Ayyâm Al-Sha'n*, and in a few passages in the *Futûhât*. We shall devote Chapter 4 to explaining the real flow of time as viewed by Ibn 'Arabî, by defining three different types of days: the normal days, and the 'taken-out' days and the 'intertwined' days.

3.4 The divine origin of the seven days of the week

Therefore, Ibn 'Arabî concludes, each Day of the seven Days of the divine Week has to be based on one of these seven basic divine Attributes:

> So because Allah (Who is described by the seven fundamental Attributes) created the world according to His own Image [II.395.25], . . . the Days had to be seven due to these (seven) Attributes and their rules, so the world appeared living, knowing, able, willing, hearing, seeing and speaking.
>
> [II.438.19]

We have already shown in section 2.1 that 'time' for Ibn 'Arabî is an imaginary attribute that is used to compare the chronological order of moving things. The day is actually a measure of the motion of the orb of the Sun, or more precisely the Isotropic Orb. And since motion is created by Allah, by creating events in the world, there are seven fundamental creative motions, or 'Days of events', each of which is originated from or ruled by one of these seven divine Attributes, and further associated with specific astronomical or astrological positions and figures apparently involved in the cosmological manifestations of those divine influences (see also the detailed outline of these associated symbols in Table 3.1 below):

> (the motion of) the Day of Sunday was from the (divine) Attribute of Hearing,
> . . . and the motion of the Day of Monday was from the Attribute of Living,
> . . . and the motion of the Day of Tuesday was from the Attribute of Seeing,
> . . . and the motion of the Day of Wednesday was from the Attribute of Willing,
> . . . and the motion of the Day of Thursday was from the Attribute of Ability,
> . . . and the motion of the Day of Friday was from the Attribute of Knowledge,
> . . . and the motion of the Day of Saturday was from the Attribute of Speaking.
>
> [II.438.7]

This relation between the seven Days and the seven main divine Attributes is not arbitrary. We notice that Sunday, which is the first day of the week (and in the creation, as we shall see in section 3.5), was from the Attribute of Hearing. This is because Allah started the creation on this day, and the first thing that is needed for the thing to be created is hearing, in order to hear Allah's creative command 'Be' [II.401.28]. Ibn 'Arabî states that everything can hear before it actually appears in the world, because everything has some sort of existence or determination in the foreknowledge of Allah before He actually creates it [II.400.7, see also section 2.3 above and section 3.6 below]. By hearing the command of Allah, the manifest world or cosmos starts to gain real existence after it had been existing in the foreknowledge of Allah. Therefore the world appears in existence as hearing, living, seeing, willing, able, knowing, and speaking; but it gains these attributes one by one in a special sequence that starts by hearing and ends by speaking – but not necessarily in the same order, owing to the 'intertwining' of those divine influences that will be explained in the next chapter.

As Ibn 'Arabî points out, we can also notice a similar sequence in the first stages of the development of the foetus and child's life: s/he acquires the first six attributes before birth, while speaking is acquired afterwards (see also section 3.6 below for more comparisons between the macrocosm of the world and microcosm of the true human being). This can be compared to the creation of the world by Allah '*Who created the Heavens and the Earth and all that is between them in six Days*' [from Sunday to Friday, per IV.11.30], '*then He mounted on the Throne*' [on Saturday, per IV.11.31] (25:59), with Saturday being associated with the divine Speaking. As Ibn 'Arabî indicates in this regard [III.108–109], Allah also says: '*The All-Merciful, (He) taught the Qur'an, (then He) created the*

Human Being, (then He) taught him speaking out' (59:1–4), which indicates again that speaking came last. On the other hand, Ibn 'Arabî affirms that the baby who is born after six lunar months may live to be quite healthy, and those six months are six lunar 'days'.

So for Ibn 'Arabî, the creation of the world, like that of the foetus, is completed in six Days: these correspond to the six directions: (up, down, right, left, front, back), and then on the Seventh Day (Saturday) it – both the world and the human being – continues living, changing from one state to another. That is why Ibn 'Arabî calls Saturday the 'Day of eternity' as we shall see further below (end of section 3.5). So like the genesis of each human being, although we do not clearly recognize it, the world – with all what it includes – 'appeared living, knowing, able, willing, hearing, seeing and speaking' [II.438.19] in seven divine Days.

Like the ancient Babylonians and Egyptians, Ibn 'Arabî also assigns a specific divine Day to each planet of the seven moving planets in the seven heavens [I.154–156]. In addition to that, he also assigns to each of these planets – following the detailed indications in certain hadith – the spiritual reality (*rûhâniyya*) of one particular prophet (Chittick 2002: 201–230). Furthermore, each one of these seven prophets always has a representative on Earth as one of the seven 'Substitutes' (*abdâl*) of the spiritual hierarchy, each one located in one region (*iqlîm*) of the seven basic geographical regions of the Earth.[12] Ibn 'Arabî always affirms that whatever happens on the Earth is preordained in the heavens day by day and hour by hour – through these spirits of the prophets and other angels who reside in the heavens – and received by their agents (*nuwwab*, s. *nâ'ib*) on the Earth, though he also affirms that there are always other unpredictable actions or events due to the direct relation between Allah and every entity in the world, which is distinct from those relations mediated through the First Intellect or the Soul [II.434.8].

Thus he explains that

> He (*the prophet Idrîs*)[13] told his friends that there are seven (spiritually perfect) Men called (the 'Substitutes' (*al-abdâl*, s. *badal*) by whom Allah keeps the seven geographical regions (*aqâlîm*); to each Substitute (is assigned) a region. They are looked after by the spirits of the seven Heavens, and each person of those (Substitutes) has power from the spirits of prophets residing in these Heavens. They (i.e. the prophets in the seven heavens) are Abraham (*al-Khalîl*, in the highest, seventh heaven of Saturn) then (in descending order) Moses, then Aaron, then Idrîs, then Joseph, then Jesus, then Adam, may Allah's peace be upon them all. As for John (the Baptist), he alternates between (the heavens of) Jesus and Aaron. So (the divine knowledge) descends upon the hearts of those seven Substitutes from the realities of those prophets – peace be upon them!; and they are looked after by the seven planets by what Allah, the Exalted, entrusted (in those planets) through their rolling in their orbs and by what Allah entrusted in the motions of these seven heavens of secrets, knowings, and higher and lower

> effects. Allah said: '*and He inspired in each heaven its mandate*' (41:12). So they have in their hearts in each hour and in each day what the possessor of this hour and the ruler of this day give away (to them).
>
> [I.154.34]

Ibn 'Arabî also shows elsewhere that the seven Days are 'created' by the corresponding seven divine Names, and those are not the same as the fundamental Names or Attributes that initiated their motion [II.442.5]. Each one of those seven creating Names also created the corresponding heaven, the prophet who is in it, as well as the planet within that orb, a specific letter from the Arabic alphabet, and one specific constellation where 'the specific planet of this orb is first created by Allah and started moving in this constellation' [II.445.5]. The relation between each heaven and the corresponding letter of the alphabet is that 'this heaven has some (special) effect in the existence of these letters' [II.445.3]. All this is summarized in Table 3.1, which is arranged from the First Day (Sunday) to the Seventh Day (Saturday).

The information brought together in this Table 3.1 provides essential keys to Ibn 'Arabî's cosmology, astrological symbolism, and understanding of the spiritual hierarchy. However here we are mainly concerned with the seven Days of the Week. It is particularly important to notice the 'irregular' relationship between the order of the days as we witness them (Sunday, Monday and so on) and the corresponding order of the seven heavens and the seven related regions on the Earth. The week starts with Sunday, from the middle (fourth, solar) orb, which is the heart of all orbs [II.275.31], then Monday in the orb of the Moon (the first, lowest heaven), then the fifth heaven, and so on. However, the corresponding order of Earth regions ('climes') goes from the fourth, seventh, third, and so on as in Table 3.1. Thus we can see that there is always a difference (separation) of three days between one heaven and the other that is directly above it. As we shall see in the coming chapter, Ibn 'Arabî therefore distinguishes between the witnessed order of the days of the week and their actual flow in the heavens, and he calls this separation *Salkh* or 'taking-out' [*Ayyâm Al-Sha'n*: 7], a rare expression taken from the Qur'an (36:37). In interpreting that Qur'anic verse, Ibn 'Arabî explains that there are three nights' and three daytimes' difference between the actual daytime and its own night that it was taken out of. This process of taking the daytimes out of the nights and vice versa is, in Ibn 'Arabî's view, the ultimate reason underlying the creation itself: for without it the world would not appear in existence in six dimensions [*Ayyâm Al-Sha'n*: 8–9]. This will be explained further in the following chapter with diagrams and illustrations.

3.5 The significance of each day of the divine week

Because of the importance of the Days of the Week of divine creation in Ibn 'Arabî's cosmology, he writes about them at length in the *Futûhât* and other books. In his mysterious book *Al-Tanazzulât Al-Mawsiliyya* (*The Spiritual*

Table 3.1 The correspondences between the seven days of the divine week and the seven heavens, seven earthly regions, divine names, lunar mansions and letters of the Arabic alphabet

Day number	*1*	*2*	*3*	*4*	*5*	*6*	*7*
Arabic Day Names	*al-ahad*	*al-ithnayn*	*al-thulâthâ'*	*al-arba'â*	*al-khamîs*	*al-jum'a*	*al-sabt*
English Day Names	Sunday	Monday	Tuesday	Wednesday	Thursday	Friday	Saturday
Divine Attributes	Hearing	Living	Seeing	Willing	Ability	Knowing	Speaking
Prophets' Names	Idrîs	Adam	Aaron	Jesus	Moses	Joseph	Abraham
Arabic 'Planet' Names	*al-shams*	*al-qamar*	*al-marrîkh*	*al-kâtib*	*al-mushtarî*	*al-zuhara*	*kaywân*
English Planet Names	Sun	Moon	Mars	Mercury	Jupiter	Venus	Saturn
Creating divine Name	*al-Nûr*	*al-Mubîn*	*al-Qâhir*	*al-Muhsi*	*al-'Alîm*	*al-Musawwir*	*al-Rabb*
Prayer	*Imam*	*Ma'mûm*	*'Ishâ*	*'Asr*	*Zuhr*	*Maghrib*	*Subh*
Constellation–Arabic	*Simâk*	*Iklîl*	*'Awwa*	*Zabana*	*Sirfa*	*Ghafr*	*Khirtan*
Constellation–English	Arcturus	Corona	Bootes	Librae	Virginis	Cover	Mane
The Heaven	fourth	first	fifth	second	sixth	third	seventh
Earthly Region/Level	fourth	seventh	third	sixth	second	fifth	first
Arabic Letter	*Nûn* (ن)	*Dâl* (د)	*Lâm* (ل)	*Ṭâ'* (ط)	*Dâd* (ض)	*Râ'* (ر)	*Yâ'* (ي)

Note
The information in this table is extracted from chapter 198 of the *Futûhât* [II.456–458] and *Al-Tanazzulât Al-Mawsiliyya*

Inspirations at Mosul, Concerning the Secrets of Prayers and the Original Days), he devoted the last ten chapters to these divine Days and their relations with the heavenly orbs and the seven planets. However, it is not easy at all to extract a lot of information from this book because he intentionally used a very difficult language full of signs and secrets. However, some of his sayings there provide very useful information when we compare them to his statements in other books. For example, right in the full title of that book, we are led to understand that there are 'original' Days, as opposed to the normal witnessed days on Earth. We shall discuss in more detail the different types of days that Ibn 'Arabî introduces and talks about there in the following Chapter 4, but what we shall now outline in the remainder of this chapter relates mostly to the 'original' Days of divine creation, which are 'the Days of (divine) Events', and not our normal witnessed days of the week.

To begin with, because each Day of the divine Week is ruled by a specific divine Name, it has to have a special significance related to this Name. This distinctive significance is not so apparent in the normal, 'circulated' days as it is for the original Days of Events as they were created by Allah. We shall first outline here the divine significance of each Day of the Week in the process of creating the macrocosm of the heavens and the Earth – and also the microcosm of the theomorphic human being (*insân*).

In addition to those general verses in Qur'an in which Allah states that He '*created the Heavens and Earth in six days and then He mounted on the Throne*' (7:54, 10:3, 11:7, 25:59, 32:4, 50:38, 57:4) there are some details about what He created each Day in the verses at the beginning of the Sura *Fussilat* (41:9–12). These verses were explained further in one hadith where the Prophet Muhammad said:

> Allah, the cherished and the glorified, created the first Day (event) in Sunday; and the earth was created on Sunday and Monday; and on Tuesday and Wednesday the mountains were created, the rivers were cleft, fruits were planted and in each earth its food was determined, *then turned He to the heaven when it was smoke, and said unto it and unto the Earth: Come both of you, willingly or unwillingly. They said: We come, obedient. So He completed them seven heavens in two Days and inspired in each heaven its mandate* (41:11–12), on Thursday and Friday; and the last of creation was in the last hours of Friday, and when it was Saturday there was no creation in it.
>
> [*Kanz*: 15120]

So here we will briefly outline what Ibn 'Arabî has to say about the significance – in the creation of both the world and the human being – of each Day of the original Days of the divine Week of creation:

3.5.1 'The first day' of creation (al-ahad, Sunday)

The Arabic name for Sunday, *al-ahad*, is also a Qur'anic name of Allah (112:1, etc.) which means 'the One', 'the Unique', the indivisible Unit. It is the first Day in which Allah started the Creation, so it is the first Day in the world [IV.11.31]. As we explained in section 2.3, the entities of the world were already eternally determined in Allah's foreknowledge, and by this state of pre-existence they gained the first fundamental divine attribute of 'hearing'. So the motion of 'Sunday was created from the Attribute of Hearing; that is why everything in the world hears the divine Command (Be! *Kun*) (even) in its state of non-existence' [II.438.8]. Allah said in the Qur'an: '*But His Command, when He intendeth a thing, is only that He saith unto it: Be, and (then) it is*' (36:82). Therefore Allah creates by saying the command 'Be' unto the things even before their actual existence. Ibn 'Arabî, however, carefully distinguishes between the divine 'Saying' (*qawl*) and 'Speaking' (*kalâm*): 'By "Saying" the non-existent hears. (This is) like His saying "*and our Command when We intend a thing is to say unto it 'Be'*" (16:40). And by "Speaking" the existent hears, as in His saying "*and Allah spoke to Moses, speaking*" (4:164)' [II.400.7].

As we have already explained in earlier chapters, Ibn 'Arabî affirms that the motion of the Isotropic Orb determines only one Day, which is one cycle starting from when the first degree of Gemini is matching the Foot in the Pedestal (see section 2.4). So because the divine Pedestal is above the Isotropic Orb, which has no distinguishing signs, the length of the day cannot be known [II.437.34]. Although we on Earth normally measure the day by hours and minutes or by the time of the Earth's motion around itself, this is a mere convention – while the actual length of the day/Day is known only to Allah [I.122.28].

Ibn 'Arabî elsewhere explains that the Sun and its heavenly orb were created on Sunday [I.155.6, I.466.6, II.445.15]. This is because the Sun resembles the spirit [I.55.8, I.275.26], and the absolute Spirit (i.e. 'the real through whom creation takes place': *SDK*: 132–134) is the first appearance of the divine Real (in the Creation), so with the initial creative motion of this First Day, the 'point' (*nukta*) or zero-dimension (0-D, as we shall discuss the dimensions in detail in section 7.10) was defined. The Sun is in the central, fourth heaven from the Earth, and this heaven was created by the self-disclosure (*SDK*: 91, 103, also *SDG*: 47–57) of the divine Name 'the Light' (*al-Nûr*). Hence:

> (the Sun) is the heart of the world and the heart of the (seven) heavens. Allah created it on Sunday, and He made it a place for the Pole of human spirits, Idrîs, peace be upon him. And Allah called this heaven '*a high place*' (19:57) because it is a heart, although the heaven that is above it is higher (in physical place). But Allah meant the highness of status (*makâna*), so the place (*makân*) (of the fourth, central heaven) is high because of its status, and Allah created it in *al-simâk* (which is the central, fourteenth station of the 28 stations or 'mansions' of the Moon), and created its planet (the Sun) and its orb, and created the letter *nûn* (ن) out of it.
>
> [II.445.15]

Ibn 'Arabî then explains the relation and correspondence between the fourth heaven (celestial sphere) of this Day and the particular geographical clime on the Earth that was appointed to the corresponding figure from among the seven Substitutes (*abdâl*), and what are the particular 'affairs' that happen on – i.e. through the higher influences associated with – this Day:

> Every affair of knowledge in the First Day is from the matter of Idrîs, peace be upon him! And every higher (celestial) effect on that Day in the elements of air and fire is from the orbiting of the Sun and its supervision which is entrusted to it by Allah, the Exalted. As for what comes from the effect on the elements of water and earth on that Day, (it) is from the motion of the fourth orb (containing the Sun). The (earthly) place of the (spiritual) person (*badal*) who upholds that influence among the climes is the fourth clime. So what is acquired, among the (divinely inspired) knowings, by this particular person among the *abdâl* (residing) in this region is the knowing of the secrets of the spiritual entities (*rûhâniyyât*: i.e. directing the heavenly spheres), the knowing of light and radiance, the knowing of the lightning and the rays (of light), and the knowing of every luminous material body – why it becomes illuminated, what is the (distinctive) constitution that gave it this receptivity (to luminosity).
>
> [I.155.6]

Also in chapter 46 of *Al-Tanazzulât Al-Mawsiliyya*, Ibn 'Arabî mentioned many mysterious details about this Day and his visit to the Pole of all spirits, the prophet Idrîs, in the orb of the Sun. This chapter (46) of the *Tanazzulât* – together with its following few chapters, and taken in conjunction with related passages from chapters 15 and 198 of the *Futûhât* – deserves a separate specialized study which is beyond the scope of this book. I have just mentioned some related passages for the 'First Day' above as an illustration of those mysterious allusions in the *Tanazzulât*, but in the shorter summaries from the *Futûhât* below will have to pass over Ibn 'Arabî's corresponding remarks for the other Days.

3.5.2 'The second day' of creation (al-ithnayn, Monday)

Ibn 'Arabî says that 'the motion of Monday was created from the (divine) Attribute of "*The Living One*" (*al-Hayy*), and through it life was in the world, so everything in the world is living (i.e. starts to become living, on Monday)' [II.438.9]. Then he explains that the Moon, which corresponds to the 'Second Day' of Creation, is in the first celestial sphere above the Earth:

> And the divine Name '*The Clarifying One*' (*al-Mubîn*) was intent on bringing into existence this lowest heaven and its planet (the Moon) on the Second Day, in the (lunar mansion) of *Iklîl* ('the Crown', which is the seventeenth station of the 28 lunar stations), and the letter *dâl* (د) is from the motion of this orb.
>
> [II.445.30]

The Moon is the fastest moving planet in the heavens, moving through one lunar mansion every day, so it goes through all 28 lunar mansions in its day, which equals 28 Earth days (according to Ibn 'Arabî, as explained in section 3.2). Ibn 'Arabî explains that from this motion the 28 letters of the alphabet are created, regardless of how they are written or spoken in different languages [II.448.4] (i.e. the sounds). Allah made this first (lowest) heaven the place for the first prophet Adam, since he is the (first manifestation or exemplar of) the Perfect Human Being.

The Moon – in keeping with its association with Adam (the emblematic human being) – resembles the soul [I.499.5], and it is the second state of the appearance of the Single Monad, after the creation of the spirits (angels), which are purely spiritual creatures. That is why Adam himself lectured Ibn 'Arabî about the directions when he visited him in the orb of the Moon (*Al-Tanazzulât Al-Mawsiliyya*: 261) – because through the creation of his Day (Monday) the space and the six directions started to form.

3.5.3 'The third day' of creation (al-thulâthâ', Tuesday)

And for the motion of Tuesday Ibn 'Arabî says that it was created from the divine Attribute 'Seeing' (*basar*): so there is no part of the world but that it is witnessing its Creator – i.e. in relation to its own individual essence, not the Essence of its Creator [II.438.10], because the Essence of Allah, the Creator, may not be seen, He may be seen only through the manifestation of His Attributes (Names) throughout the Creation as we shall discuss in more detail in section 5.5. Ibn 'Arabî then adds that

> The divine Name 'the All-Prevailing' (*al-Qâhir*) was intent on bringing into being the third heaven (the fifth from Earth); so He caused its (distinctive spiritual) reality to appear, along with its planet (Mars, *al-marrîkh*) and its sphere, and He made it the dwelling for Aaron.... The existence of this planet and the motion of its sphere were in the lunar mansion *'Awwa* (which is the thirteenth station of the 28 Moon stations) on Tuesday.... And from the motion of this sphere appeared the letter '*lâm*' (ل).
>
> [II.445.8]

3.5.4 'The fourth day' of creation (al-arba'â', Wednesday)

For Wednesday he says that 'the motion of the Fourth Day came into existence from the (divine) Attribute of Willing (*al-irâda*), so there is no part of the world but that it is seeking to glorify the One Who gives it existence [II.438.11].

Mercury is in the sixth sphere from the Earth, and this heaven was created through the (self-disclosure of the) divine Name 'the Enumerator' (*al-Muhsî*). Allah created this heaven, its planet (Mercury), the Fourth Day (Wednesday), and letter *tâ'* (ط) in the lunar mansion of the constellation *Zabana* (which is the sixteenth station of the 28 Moon stations), and He caused Jesus to dwell there [paraphrasing II.445.26].

3.5.5 'The fifth day' of creation (al-khamîs, Thursday)

And the motion of the Fifth Day (Thursday) came into existence from the (divine) Attribute of Ability (or 'Power', *al-qudra*), so there is no part of existence but that it has been enabled to praise the One Who gives it existence [II.438.16].

Jupiter (*al-mushtarî*) is in the sixth heaven from the Earth, and it and its sphere were brought into existence through the self-disclosure of divine Name 'the All-Knowing' (*al-'Alîm*). Allah created this heaven, its planet, the Fifth Day and the letter *dâd* (ض) in the lunar mansion of the constellation *Sirfa* (which is the twelfth station of the 28 Moon stations), and He made it a dwelling place for the prophet Moses [paraphrasing II.444.25].

3.5.6 The sixth 'day of gathering' (al-jum'a, Friday) and its special hour

Friday holds a special importance for Muslims and especially for Ibn 'Arabî, who says that in this Day our Tablet (i.e. the souls) received its secrets from its Lord, through the pens (i.e. the intellects). Ibn 'Arabî then says that 'the motion of the Day of Gathering (Friday) came into being through the (divine) Attribute of Knowledge: so there is no part of the world but that it knows the One Who gives it existence, with regard to its own essence, not the Essence of the One Who gives it existence' [II.438.13].

Venus is in the second sphere of the heavens from the Earth and this heaven was brought into existence by the self-disclosure of the divine Name 'the Shaper' (*al-Musawwir*, the One Who gives form). Allah created this heaven, its planet (Venus), the Day of Gathering (Friday), and the letter *râ'* (ر) in the lunar mansion of the constellation *Gafr* (which is the fifteenth station of the 28 stations of the Moon), and He made it a dwelling place for the prophet Joseph [II.445.23].

Before the advent of Islam, the Arabs called Friday *'Urûba* ('the day of beauty and adornment') [I.645.25], because it was a distinguished day in which people meet. But in Islam Friday is called *al-jum'a* ('the Day of Gathering') because that is the day people gather in prayer in one mosque. Ibn 'Arabî, however, gives another explanation of the deeper ontological meaning for *Jum'a*. On this Day Allah created the Human Being (Adam) in His Image, so on this Day Allah 'gathered together' (Arabic verb *jama'a*) the form of the Truly Real (*al-Haqq*) and the form of all creation in the theomorphic human being (*insân*), so that is why this Day is called *Jum'a* [I.643.27]. And this special divine 'gathering' of the theomorphic human being has occurred on a specific hour of this Day, which made this whole Day holy, noble and the best of all Days.

Ibn 'Arabî affirms that Allah created humanity (Adam) on Friday [I.466.15], following a well-known hadith in which the Prophet Muhammad said: 'The best Day the Sun rises on is Friday: Adam was created in it, he entered into Paradise in it, he was taken out of Paradise in it, and the Hour (of the Resurrection) will only come on Friday' [*Kanz*: 21050].

Ibn 'Arabî goes on to explain, on the basis of another hadith [*Kanz*: 21063], that some people argued about the best day of the week to celebrate or commemorate God in, and that the people of the three major religions each followed a different way, especially since Allah did not specify which day is best, but left it for them to discover:

> So the Christians said the best of days – and Allah knows best – is the First Day (Sunday), because it is the day of the Sun and it is the First Day in which Allah created the heavens and the Earth and what is between them. He started the Creation in it due to its pride over the rest of the days. So they took it as a festival.... And the Jews said: 'But it is Saturday, because Allah completed the Creation on Friday and He rested on Saturday.' ....
>
> Then for this community (of Islam) Gabriel came to Muhammad, may Allah have peace and mercy on him, with the Day of *Jum'a* in the form of a polished mirror in which there is a spot (*nukta*), and he said unto him: 'this is the Day of *Jum'a* and this spot is an hour in which no Muslim servant happens to be praying (when it comes to pass), but Allah shall forgive him' [*Kanz*: 21063]. So the Prophet Muhammad, may Allah's peace and mercy be upon him, said: 'So Allah guided us to what the people of the Book disagree about, which is the divine explanation with the mirror' – and he assigned the guidance to Allah. And the reason behind its superiority is that it is the Day in which Allah created this human kind, for whom He created the creatures from Sunday to Thursday, so it has to be the best of times. And the creation (of human kind) was in this Hour which appeared as a spot in the mirror.... And this Hour in the Day of *Jum'a* is like the night of destiny [*laylat al-qadr*, which according to Qur'an, *is better than a thousand month*s (97:3)] in the year.
>
> [I.466.6]

So, he concludes, that is why Allah chose this Day over the other Days:

> As He, the Exalted, did with all the existents, He has chosen as the best one thing from each kind. He has chosen as the best from the Beautiful Names the Name '*Allâh*'; He has chosen as the best from the people the prophets; He has chosen as the best from His servants the angels; He has chosen as the best from the heavenly spheres the (divine) Throne; He has chosen as the best from the elements water; He has chosen as the best from the months Ramadan; He has chosen as the best from worshipping fasting; He has chosen as the best from the centuries the century of the Prophet Muhammad, may Allah have peace and mercy upon him; He has chosen as best from the Days the Day of Gathering (*al-jum'a*); He has chosen as best from the nights the Night of Destiny; He has chosen as best from (good) deeds the prescribed deeds (*al-farâ'id*); He has chosen as best from the numbers the number 99 (in the hadith on the 99 'Most Beautiful Names' as mentioned above).
>
> [II.169.5]

Then Ibn 'Arabî goes on to explain why Allah has chosen as best these particular things in each kind. With regard to the Day of Gathering, he says:

> As for (the reason) for His choosing Friday as best from the Days (of the divine creative Week), that is because in it the Two Forms appeared [i.e. of the cosmos and of the theomorphic human being, in their final perfected state]. Allah made that Day (of the Gathering) for the form . . ., and it is (uniquely) a feminine Day, for which is the beautiful adornment and the completion of the creation (through Adam).
>
> And [alluding to the hadith just cited above] Allah has chosen an Hour of its hours that is like a spot in the mirror. That is the place (of manifestation for) the form of the one [i.e. Adam, as the emblematic fully human being and divine vice-gerent] who is disclosed in the mirror of this Day. Thus He/he sees in it/him H/himself.[14] And through that Form [of all human being] that appears *between* the mirror (of creation) and the One Who looks in it, there takes place the (divine) Addressing (of each human soul) and the imposition of responsibility (*taklîf*: on each human being).
>
> [II.173.15]

So this Hour is the most precious hour, because in it the divine Form of the Real (*al-Haqq*) and the form of the theomorphic human being meet together (*yajtami'ân* from the same Arabic root as *al-jum'a*), and by this 'gathering' the Real is manifested – and at last perceived – in the Creation in the most perfect form, which is the form of the perfect human being. In another passage, Ibn 'Arabî explains this further:

> And there is no Day among the Days (of Creation) more perfect than the Day of *Jum'a*, because in it there is made manifest the Wisdom of (God's) Capability, through His creating on that Day the human being whom Allah created on His Image. So there remained no (more) perfection for the divine Capability to create, since nothing is more perfect than the Image of the Real (*sûrat al-Haqq*). . . .
>
> Now since God specially distinguished (this Day) by that 'Hour' which does not belong to any other Day – and time (*zamân*) is nothing but those Days – therefore this Hour does not pertain to any other times but to the Day of *Jum'a*. It is one part of the 24 parts of the Day, and it is in the half of the Day which is called daytime (*nahâr*). So it is in the manifest dimension (*zâhir*) of that Day, and in the inner dimension (*bâtin*) of human being, because the outer dimension of human being corresponds to the inner dimension (night-time) of the Day, while the inner dimension of human being corresponds to the outer dimension (daytime) of the Day.
>
> [I.645.26]

Ibn 'Arabî also comments on the imperative question of whether this Hour is fixed, or is floating all over the Day, so that each Week it comes at a different time, since the original hadith above [*Kanz*: 21063] did not give any explicit indication of its exact location 'in' the Day of Gathering. Ibn 'Arabî, in any case, does not rule either possibility. He suggests that, if we have to take the strict sensational meaning of the analogy between the Day and the mirror and the forms that appear in them, then that specially chosen Hour would be fixed. But, if we want to take it as an abstract meaning in the imagination, without moving it into the world of senses – which is quite plausible in this context – then we could say that it may be floating all over the Day of Gathering. So he affirms that both possibilities are plausible and that the issue may not be resolved without a further divine specification [I.466.17]. However, because it denotes the creation of the Human Being as the Image of the Real, this Hour must be fixed (on the Day of Gathering) among the original (divine) Days that are the 'Days of events', although it might be floating in the normal 'circulated days' that we actually experience, because they are 'intertwined' with the 'Days of events'.

3.5.7 The seventh 'day of rest' (al-sabt, Saturday) as the 'day of eternity'

As for the last day, Ibn 'Arabî says that 'the motion of Saturday was created from the (divine) Attribute of Speaking (*kalâm*), so everything in the existence glorifies in thanks of its Creator, but we do not understand their glorification' [II.438.16]. Obviously Ibn 'Arabî is referring here to the verse '*the seven heavens and the Earth, and all beings therein, declare His glory: there is not a thing but celebrates His praise; and yet ye understand not how they declare His glory! Verily He is Oft-Forbearing, Most Forgiving!*' (17:44). We shall also see in section 7.8 that the world essentially is the 'words' (i.e. the vibrations) generated by its entities on the smallest scale. This is true for whatever we hear, see, smell, taste and everything physical – all are the 'words' spoken by the entities of the world. This seems to cope very well in accordance with the recent theory of Superstrings.

Elsewhere Ibn 'Arabî says:

> Saturn (*Kaywân*) is in the seventh sphere from the Earth, and this heaven was created by the (self-disclosure of the) divine Name 'the Lord' (*al-Rabb*). Allah created this heaven, its planet and the 'Day of Rest' (*al-sabt*, Saturday) in the lunar mansion of the constellation *Khirtân* (also called *al-Zabra*, and it is the eleventh mansion of the 28 Moon mansions), and He made it a dwelling place for the prophet Abraham.
>
> [paraphrasing II.442.21]

Saturday bears a very important and unique meaning for Ibn 'Arabî, since he considers it to be the Day of eternity. This concept is initially mysterious, but extremely important, since Ibn 'Arabî mentions it many times in the *Futûhât* and other books and considers it as an undisputable fact.

In his *Al-Tanazzulât Al-Mawsiliyya*, he asserts that: 'Saturday passes through the existent things like number does in the countable things, permanence in the permanent things, and standing in the standing things; it is not non-existing nor existing, and not present nor absent' (*Al-Tanazzulât Al-Mawsiliyya*: 339).

This description is also like the strange properties of the 'Universal Reality' (*haqîqat al-haqâ'iq*, 'the Reality of Realities') that he introduces to connect the divine properties of the Real with the properties of the Creation. By way of clarification, he then goes on to explain that

> The whole world (as kinds and species, including metals, elements, plants, animals, angels, jinn and humans) from its beginning to its end was formed in six Days, from the beginning of (the night) of the First Day until the end of the Day of Gathering (Friday), and there only remains Saturday for changing from one state to another and from one rank to another, and for transmutation from one being to another, constantly. . . . That is why the 'rulers' (in terms of higher astrological or cosmological influences) on this Day are cold and dryness (so that they will be able to hold the picture in imagination in order to feel the continuous presence, as will be discussed in section 5.6), and among the planets: Saturn. So this Day alone was a circumferential orb whoever moves in it will know the Identity, the Attributes, the Actions and the Effects.
>
> [I.61.13]

This alludes to the three types of divine Names: the Names of Essence (*asmâ' al-dhât*), the Names of Descriptions or the Attributes (*asmâ' al-sifât*) and the Names of Actions (*asmâ' al-af'âl*) [I.423.20, I.67.28; *K. al-Masâ'il*: 97] and the 'Effects' here are the entire Creation. So Ibn 'Arabî says that the one who moves with the orb of Saturday will know the Real and the Creation. This is in fact the description of the spiritual 'Pole' – like Ibn 'Arabî himself – who is therefore 'out of time' (see also section 2.7), because he already knows that he witnesses the real Day of Saturday.

Ibn 'Arabî even explicitly calls Saturday the 'Day of Eternity' (*yawm al-abad*: *al-abad* being eternity considered as extending eternally 'into the future', as opposed to *al-azal* or 'beginningless eternity') quite often in his writing. For example, in the *Futûhât* he says:

> And so the Day of Eternity starts which is Saturday (*al-sabt*), and *al-sabt* means 'rest'; it is the seventh Day, and it has no end. No sense of weariness touched the Creator in creating what He did. So Saturday was the Day of being finished with the creation of the levels of the world, but there still remains for Allah the creation of what this world requires of the states (for every individual creature), which are eternally endless and without limit.
>
> [III.192.20]

He also named chapter 53 of *Al-Tanazzulât Al-Mawsiliyya* 'On the Inner (Spiritual) Knowing that Saturday is the Day of Eternity, and it is the Day of Transmutations'. Saturday, therefore is the 'Day of Rest' (as its name in Arabic, *al-sabt*, also indicates), because Allah has finished the creation (of all the cosmos and human being) on Friday, so that there remained nothing new to be created – except for the endlessly changing states of all the creatures in time.

Thus Ibn 'Arabî affirms that

> Allah created the world in six Days; He started on Sunday and finished by Friday, and no sense of any exhaustion touched Him; nor did He get weary from creating the creatures. So when it was the seventh Day of the week and He had finished creating the world, He looked like someone who is taking a rest from weariness. So He reclined and put one leg over the other and said: 'I am the King', as narrated in the prophetic sayings, so this Day is called *al-sabt*, which means 'the Day of Rest'. It is the Day of Eternity, because in this Day the individuals of every species are formed in the world and in the hereafter.
>
> There are only seven Days, and each Day has a ruler (*wâli*) designated by Allah. So when the matter/Command ended up on Saturday, Allah assigned it to a ruler with the ability to stabilize and fix, so he can stabilize the forms (of all the creatures) in the '*Dust*' (*al-habâ'*, of the *barzakh* or divine Imagination). So the daytime (*nahâr*) of this Day – i.e. the Day of Eternity – is for the people of the Gardens, and its night-time is for the people of the Fire: so there is no evening (end) for its daytime, and its night-time has no morning.
>
> [IV.11.30]

Then Ibn 'Arabî mentions the story of his encounter with the (spirit of) *al-Sabtî* (lit.: 'the Saturday man') who is Muhammad, son of the famous Abbasid caliph Harûn al-Rashîd, whom he met in Mecca one Friday while circumambulating the Kaaba, and asked him: 'I knew that you were called *al-Sabtî* because you worked every Saturday for what you ate during the rest of the week . . . so why did you choose Saturday from the rest of the days of the week?' Al-Sabtî replied:

> This is a very good question. I knew that Allah started creating the world on Sunday and finished on Friday, and on Saturday He reclined and put one leg over the other and said: 'I am the King'; that is what I heard in narrations when I was in this world. So I swore to work like that.
>
> [IV.11.35, see also *Al-Tanazzulât Al-Mawsiliyya*: 340]

So he worked on Saturday and rested (to devote himself to worship Allah) in the other six days. So Allah's work of the creation was done in the six Days from Sunday to Friday, and the 'Day of Eternity' is for us to work and develop through the path leading to our ultimate destination.

On the other hand, the fact that the seventh Day was created through the divine Attribute of Speaking reminds us that the very First Day was created through the divine Attribute of Hearing. Ibn 'Arabî typically points out that this all together is like a circle: since the world is created through divine Speaking (the Command 'Be'), so the creatures first needed hearing in order to hear that existentiating Command. Because human beings are – at least potentially – uniquely created in the fully theomorphic Image of Allah, He granted them something of the Attribute by which He creates things, so they in turn can create through 'speaking': these are, for example, the meanings one creates in the soul of the hearer when one speaks. And because creation is in the end a divine act, the Prophet Muhammad says that 'Allah is by the tongue of every speaker' [*Kanz*: 7842; see also IV.187.19, IV.292.32]. So this brings us back to the initial receptive state of 'Hearing', because everything in the world is at the end perceived through vibrations, even the visible and other things. Therefore, the creation takes place always, at every moment in time and in every point of space. In terms of the same divine creative cycle of the seven Days explained above, the world is re-created every moment in six Days (from Sunday to Friday) and displayed on the seventh Day (Saturday). Indeed in terms of Ibn 'Arabî's famous theory of the oneness of being, the speaker is in fact Allah, and the hearer is also nothing other than Allah [II.367.16]. Ibn 'Arabî frequently expresses this paradoxical reality in language referring to the forms of the ritual prayer, as when he says: 'Thus Allah says by the tongue of his servant (as he says during the prayer) 'Allah hears who praised Him'; and then the servant says 'Our Lord, to Thee is the Praise!', not to me' [II.367.14]. Also he concludes: 'You praise Yourself (on the tongue of every speaker), and You hear Your words. Because *the Praise is* only *for Allah and by Allah* [I.112.32, alluding to Qur'an 1:1].

One final aspect of Saturday is that Ibn 'Arabî also describes it as 'the fruitless Day' (*al-yawm al-'aqîm*; 22:55), which is a result of its being the Day of Eternity:

> Now the fruitless is that from which nothing is born, so it does not have a birth of its kind. So (Saturday) is called 'fruitless' because at the end there is no Day after it, and this refers to Saturday from the (divine creative) Week, which is the Day of eternity. Its daytime is all light for the people of the Gardens eternally, and its night-time is all darkness for the people of Gehenna, also eternally.
>
> [III.564.22]

3.6 Space-time and creation in six days

So this means that we – along with all of manifest creation, at all times – are now living in this 'Seventh Day', but the riddle is how we can at the same time still be viewing the other Days of the week, including Saturday itself. Ibn 'Arabî himself remarks that 'it is rather amazing that the Days, among which is Saturday, are (all) happening in Saturday; because it is one of these Days and they are appearing in it' [II.444.7]. But he says that this is ultimately plausible, and he gives the striking example from the famous divine saying describing Adam when Allah showed him his progeny, including himself, in His Hand – while at the same time he was outside looking on at that [II.444.14; *Kanz*: 15123].

The fact that Saturday is the 'Day of eternity' suggests that the *time* that we live is only Saturday, so that the other six Days (of events) of the (original) Week could account for space, and not time; this is also supported by the observation that the usual word for 'day' in Arabic (*yawm*) also bears the meaning of 'direction'.[15] If we add this to what we have just said above about the oneness of being, and that the act of creation 'in one Week' actually takes place at every moment of time, then we can easily come to a very important conclusion that the actual Week and its seven distinctive Days make up a unit of space-time, and not time alone. In the first six Days, space is created with its six directions (three dimensions) and then it is displayed ('in time') on Saturday. We do not feel the creation in the six Days because we are being created but not yet, and we only feel ourselves created on Saturday, which is why it is the Day of eternity. As Allah said in the Qur'an: '*I called them not to witness the creation of Heavens and the Earth*' (18:51), which He created in six Days, from Sunday to Friday. But when this initial stage of creation is over and Allah the All-Merciful '*mounted on the Throne*', changing the creation from one state into another, then and only then, on Saturday, do we realize ourselves and also realize the world (of space) that was created in those preceding six Days.

Ibn 'Arabî declares in many passages that the world is constantly being re-created on every single 'Day of event' (see section 5.6). And the Days of events are seven, based on the central divine Attributes of Allah. So the creation of the world by Allah is a seven-step process: in each Day of those seven Days of events the world gains the qualities and manifestations of the corresponding divine Attribute. Thus on the First Day, the world (which had already been determined eternally in the foreknowledge of Allah) hears Allah's command: 'Be', so it moves from the 'absolute Unseen' into the first step of existence which has no dimension yet (0-D), like the geometrical point (*nuqta*); it is the first real Image of the Real (*al-Haqq*), Allah, the Exalted, and thus it is called '*the real through whom creation takes place*' (*SDK*: 132), and this is the 'Greatest Element' (see sections V6.2 and 6.5). So 'the Real' is a name of Allah and this Greatest Element. Then, on the Second Day, the world becomes living, and by that the creation of angels from the divine Light takes place (1-D), as will be discussed in more detail in section 7.10. Then on the Third Day it can witness its Creator (with respect to its being His creation), while on the Fourth Day it becomes willing to magnify[16] its Creator (and by that the creation of the jinn takes place), and on the Fifth Day it is able to do so. Finally, on Friday – or by the 'last three hours' of Friday, when the Human Being is created[17] – the world will actually know its Creator (again with respect to His act of creation, not in Himself): and this happens only by creating the Human Being (Adam), who is in respect to the world the Spirit is for the true Human Being [I.118.8, III.363.3].

By these six Days the world is completely created as a Single Monad who is also the 'Complete Human Being', including heavens and Earth, mind and soul, spirit and body. Then on the seventh Day this picture of the complete world will be held in the imagination of the human being, because it is an imaginal form that will intrinsically cease to exist right in the second moment ('Day') after its creation, to be instantly re-created in a new form. That is why, Ibn 'Arabî explains, Allah appointed as the 'rulers' on Saturday the essential qualities of coldness and dryness, so that this cosmic image may appear to be continuous [I.61.15]. So our imagination holds a new picture of the world, every 'Day' (moment), and by comparing between those succeeding pictures, the imagination perceives the motion or change in space and time, since Allah never manifests in the same form twice or to two perceivers [I.679.7, II.77.27, II.616.3]. This is also the meaning of the verse: '*each Day He is in a* (different) *task*' (55:29), which Ibn 'Arabî uses more frequently than any others from the Qur'an to corroborate this particular ontological view.

It has to be noticed however, that, in every Week of Creation, only one 'point' of space-time is created, and that is the point where the observer is – or rather, the observer himself is also created at that point in space-time. The observer perceives other points of space and time because they have been held somehow within the 'Dust' (*habâ'*) of the divine cosmic Imagination (*barzakh*, *khayâl*, etc.).

Therefore, Ibn 'Arabî explains the verse in Qur'an: '*Your Lord is Allah Who created the heavens and the Earth in six Days, then He mounted on the Throne; He draweth the night as a veil over the Day*, (7:54, see also other similar verses in Qur'an: 10:3, 11:7, 25:59, 32:4, 50:38 and 57:4) as indicating that the creation has been completed in six Days, from Sunday to Friday, by creating Adam (the cosmic Human Being) on Friday; and then '*He established Himself on the Throne*' on Saturday when nothing new is to be created, but only changing the states of the creatures – until the Day of Judgement, and after that either in the Gardens or in Gehenna – so this is the Day of eternity. Yet the heavens and the Earth are constantly being *re-created* every moment: therefore the same Week and Days of creation, including Saturday, are continually reiterating throughout Saturday, at every moment.

So if we think of the whole of manifest creation (the cosmos or 'world') throughout all places and times, throughout 'the Age' (*al-Dahr*), it is only seven Days: six Days for the creation, and the seventh Day (Saturday) for the realization. But if we think of ourselves as points in this space-time of the Age, because of the eternally renewed re-creation we observe succeeding Days and Weeks (as moments); yet all those Days (including Saturday) are happening in the Saturday of the single, unique Week of 'the Age'.

Ibn 'Arabî draws interesting symbolic parallels and extrapolations from this concept of creation just explained. Just like this momentary creation in the Week, our own human age (life) is also seven days (or periods): six in this world and Saturday for the hereafter, which we have either as all-day in Paradise or as all-night in Gehenna. Similarly also, the age of human civilization on Earth is seven days and we are now living in Friday, indeed in the last few hours of it. This is because, as Ibn 'Arabî observes:

> We (the true followers of the Prophet) are, thanks to Allah, the Day of *Jum'a*, and Muhammad – may Allah have peace and mercy upon him – is this very '*Hour*' itself, by which the Day of *Jum'a* takes precedence over all other Days. [I.646.15]

This is because Muhammad is the Perfect Human Being who is the most perfect Image of the Real. Ibn 'Arabî explains this well-known central conception in his thought in more detail in chapter 346 of the *Futûhât*, where he observes that Muhammad, in relation to the whole world (cosmos), like the spirit (*al-nafs al-nâtiqa*) for each human being, while other prophets are like other spiritual faculties [III.186.31] (see also Table 3.1). Then he indicates that the world before the appearance of the Prophet Muhammad was like a foetus in the womb of the mother [III.187.6], and that after the passing away of Muhammad the world is like a sleeping – not dead – person [III.187.1], until, in the next world, all the world shall 'live' (have real life and awakening) through the full appearance of Muhammad in his complete composition [III.187.8]. Thus he goes on to note that people in the Gardens and Gehenna are like the different (spiritual and material) parts of the earthly human being, where the people of Gehenna are like the relatively lifeless (yet not completely dead) hair and nails [III.187.10]. So in this way the world is a like an 'immense human being' (*insân kabîr*), while the fully human being is like a microcosm (*âlam saghîr*).[18]

Many earlier Muslim theologians have tried to explain that Allah did not mean by the Qur'anic references to the 'Days' of creation those same earthly days that we live in, because they claimed that the sort of days that we know could not be existing yet, prior to the completion of creation. In contrast, Ibn 'Arabî stresses that those 'Days' of creation in reality coincide with these same observed and experienced days. At the same time, he also acknowledged that this (cosmic) 'Day' existed before the creation of the heavens and the Earth, while the (earthly, observed alternation of) daytime and night-time were only determined afterwards by the creation of the Sun.

On another level, Ibn 'Arabî's complex, unique interpretation of those many verses in the Qur'an (and the Bible) that talk about the creation in six Days is promising in terms of modern astronomy and cosmology, because it suggests the first real unification between space and time, as well as a sense in which the 'Week' may have a very important, and not simply conventional, real significance.

However, there are still some further issues that need to be settled in order to understand the meaning of the Week, its movement through the signs of the zodiac and its detailed role in the process or stages of creation. We shall devote Chapter 4 to explaining Ibn 'Arabî's complicated theory of the 'Actual Flow of Time' and Chapter 6 to outlining a complete model of the cosmos based on Ibn 'Arabî's unique view of time and his theory of the oneness of being, which will be discussed in Chapter 5.

4 The actual flow of time

As we already explained, Ibn 'Arabî showed that the creation of the world by Allah is a 'series' process. That is, He creates only one event at a time (or indeed in each Day), as Allah says: *each Day He is upon some* (one) *task* (55:29). Yet we observe multitudes of different events happening apparently at once. According to Ibn 'Arabî, the way to understand this apparent paradox is to correct our view of the flow of time.

Ibn 'Arabî shows that the actual flow of time (the divine creative 'Week' as the basic unit of space-time) is not the witnessed flow from Sunday to Monday and so on to Saturday. These normal days of the week that we witness are the '*circulated*' days, and not the actual divine Days (the 'Days of Event') in which only one single event should be happening each and every single Day. In order to define or construct the actual cosmic 'Days of events', we need first to attach each night-time to its own daytime, for again the underlying metaphysical connection is not as we observe: i.e. the ('real') night-time of each daytime is not the observed night-time that precedes or follows this daytime, but rather it is '*taken out*' and separated from its own daytime by three daytimes and three night-times. This extraordinary connection, Ibn 'Arabî explains, is symbolically related to our three-dimensional structure as human beings. Finally, he also affirms (according to Qur'anic indications), that the actual Days of events are '*intertwined*' with the observable, 'circulated' days in a specific manner that he details in his book *Kitâb Ayyâm Al-Sha'n* (*The Book of the Days of Task*).

So we shall first explain in this chapter the meanings and the relations between those three metaphysically distinct types of days: the 'circulated' days (*ayyâm al-takwîr*), the taken-out days (*ayyâm al-salkh*) and the intertwined days (*ayyâm al-ilâj*). There we shall see that, on the basis of the hypothesis of 'intertwining', Ibn 'Arabî introduces a unique new concept of time (and space) that has apparently not been discussed before. Finally, we turn to explaining the distinctive motion of each Day in the zodiac and the kinds of effects that are caused in the world in these Days. And at the end of this chapter we shall see the divine origin of dividing the normal days into hours, minutes and seconds.

4.1 The original days of event

We have explained in sections 2.15–16 that there is a minimum 'Day' which is that indivisible duration of time called 'the Single Day' (*al-yawm al-fard*). And we showed that this Day – when taken globally, or indeed universally (since it encompasses all of manifest creation) – actually includes our normal day, though we encounter it in fact just as a moment. In each one of these moments/Days there is only one global cosmic 'event' happening in each part of the entire cosmos. These Single Days are the original cosmic Days which are called the divine 'Days of event' (*ayyâm-ul-sha'n*), as discussed in the preceding chapter and Chapter 2. For in each Day of event (or divine 'task') Allah creates a single event actually encompassing every entity and phenomenon of the whole manifest world. As He said in the Qur'an: '*each Day He is upon* a (single) *task*' (55:29).

Ibn 'Arabî points out that 'in each day of our normal days, that is from sunset to sunset or from sunrise to sunrise, there is the end of 360 days' [*Ayyâm Al-Sha'n*: 6] (assuming the circle is 360 degrees).[1] That is because in every moment of the normal day that there is the end (or the beginning) of a day in one place, there is a corresponding beginning (or end) of another day in another place.

To explain this further, let us divide the circumference of the Earth into 360 longitudinal lines and the day into 360 degrees of longitude. Therefore, any whole day that we encounter in any specific place is a combination or the sum of the ends (last degrees) of 360 days from other places on the Earth; or in the same way, it is the sum of the first degrees of 360 days around the Earth. For example let us suppose that we are on the first longitudinal line at the first degree of the day, then the second degree of this day on this first longitudinal line is the first degree of the day beginning on the second line, and so on. In another way we can also say that the 360 degrees of the 'day' on the first line are the collection of the last degrees of the lines 360, 359, 358 and so on up to line number 1. Therefore in every moment there is one full day around the Earth: now, for example, it is morning somewhere, noon somewhere else and evening and midnight in other places; but all in total is a single full day (see also section 2.15 above).

So the flow of these original Days marks the actual sequence of events that spans space and time. But, because we live in and can only observe a tiny point of the whole space of the globe, we encounter linear time as our normal observed days of the week (i.e. the 'circulated days' described in the next section). Therefore, these original 'Days of event' are intertwined with our normal observed days.

4.2 The 'circulated' days (*ayyâm al-takwîr*)

Then if we stick to one place and watch the flow of time, we see that the Sun sets, for example, at the beginning of 'Sunday night' (i.e. the night that begins before Sunday daytime);[2] and then it rises on Sunday morning, and so on until it completes a full week on the following Sunday night. This is the witnessed week, and its days are the normal days, or the witnessed days. Ibn 'Arabî calls them the 'circulated days' (*ayyâm-ul-takwîr*),[3] because they (daytimes and nights), seem to run after each other in a circle, as Allah says in the Qur'an:

> He created the Heavens and the Earth through the real. He causes the night-time to encircle (v. *yukawwiru*) the daytime, and the daytime to encircle the night-time. He has subjected the Sun and the Moon: each one follows a (designated) course for a time appointed. Is He not the All-Mighty, the All-Forgiving?
>
> (39:5)

So we can also say that the daytimes are circulated around the nights, and the nights are circulated around the daytimes, because the daytime and the night are like hemispheres surrounding the Earth all the time, but running after each other as if they were seeking each other, as Allah said in THE Qur'an: '*Lo! Your Lord is Allah Who created the Heavens and the Earth in six Days then mounted He on the Throne. He covers (yugshî) the night-time with the daytime, which is in haste to seek it*' (7:54).

The Arabic word *yugshî* in this verse not only means 'to cover' but also 'to embrace' [*Ayyâm Al-Sha'n*: 7], and it is particularly used for marriage. Thus Ibn 'Arabî suggests that it is as if the daytime and the night are seeking each other because they want to 'marry' each other, to produce children – since we and everything else in the world are all the 'progeny' of the daytimes and nights [*Ayyâm Al-Sha'n*: 7]. This 'abstract marriage', is a basic concept in Ibn 'Arabî's cosmological teachings, and we shall come back to it shortly.

We conclude that in every normal day (i.e. circulated day) as we observe it in one specific place on the Earth, a vast number of 'Days of events' happen, as many as there are indivisible moments in this day. In other words, in every moment that we encounter on the Earth, there is a Day of event happening that encompasses the whole world or cosmos (all of creation). Yet each Day of event is also composed of the seven distinctive Days of creation, the divine creative 'Week' described in detail in the previous chapter.

4.3 The 'taken-out' days

Allah said in the Qur'an: '*A token unto them is the night-time: We take the daytime out of it, and lo, they are in darkness*' (36:37). Ibn 'Arabî points out that this seems to indicate that night is the origin, and that daytime was somehow 'hidden' in it and then was taken out of it [*Ayyâm Al-Sha'n*: 9, II.647.20]. In other words, as Ibn 'Arabî explains [I.716.15], the night is like a dress or a skin over the daytime, and then Allah takes the daytime out of the night so that the world – which was in the absolute darkness of the divine 'Unseen' (*al-ghayb*) – is created (i.e. so that it appears in the light of actual existence).

Ibn 'Arabî, however, argues that Allah did not specify in this verse which daytime was taken out of which night, and so this has to be clarified. For it is not, as we might think, that each daytime (that we witness) was taken out of its own night. We have to seek the *true* relation between each daytime and its night, and this relation, Ibn 'Arabî says [II.445.32, III.203.30], is based on the first hour of the daytime and the night-time, because each hour of the daytime and the night-time has a ruler; one of the five planets, the Sun, or the Moon (corresponding to one of the seven principal divine Attributes: see section 3.4); so each day is named after the planet that rules the first hour of it. For example: the first hour of Sunday is ruled by the Sun, and that is why it is so named (in English and many other languages); likewise Monday is the day of the Moon, and so on. In Arabic, however, the names of the days of the week do not have direct relations with the names of the planets that rule these days, but this connection still forms a basic principle in Ibn 'Arabî's view of time.

But before we discuss this further and assign each night to its actual daytime, we should understand the exact meaning of 'taking out' the daytime from the night, or the night from the daytime. Ibn 'Arabî regards the different daytimes and night-times as 'parents' to what Allah creates in them: so everything that happens in the daytime is like a 'son' whose father is the night and whose mother is the daytime; and everything that happens in the night is like a son whose father is the daytime and whose mother is the night [*Ayyâm Al-Sha'n*: 7; II.445.18]. As Allah said in the Qur'an: '*He merges* (*yûliju*) *the night into the daytime, and He merges the daytime into the night*' (57:6). So there is a kind of abstract, generative 'marriage' between daytimes and nights, but where nights and daytimes exchange their parental roles from being fathers to being mothers, and vice versa. That is why they are 'intertwined', as we shall see further below. Now Ibn 'Arabî explains that when the daytime turns from being father into being mother or vice versa, this is what is meant by the Qur'anic reference to its respective 'stripping-out' or 'taking-out' (*salkh*). So when we say that this daytime (*nahâr*) is taken out of that night-time (*layl*), it means that this daytime and night exchange their generative 'parental' roles, although together they are always like a couple, i.e. a single 'day' (*yawm*).

Ibn 'Arabî adds that Allah did not explicitly mention that night is also taken out of the daytime, since it is readily understood from the same Qur'anic verse [*Ayyâm Al-Sha'n*: 8; I.141.6, I.716.11]. On the other hand, Ibn 'Arabî indicates that the first hour of the daytime and that of its own night (which was taken out of it) should be ruled by the same planet [*Ayyâm Al-Sha'n*: 10]. For in order to consider the daytime and the night-time as one unitary day, they have to be ruled by the same (cosmological, planetary) 'ruler'. This applies only to the first hour of the daytime and the night, because each of the other hours (of the observable, earthly days) is coming from other cosmic 'Days' as a result of their overall 'intertwining':

> Now when these planets moved in their orbs, Allah made for each planet a (specific) Day among the Days of the zodiac-orb motion.... So He defined for each Day (*yawm*) a daytime (*nahâr*) and a night-time (*layl*), and He distinguished between each night and its daytime by the rule of the (particular) planet for that Day in which the daytime and night appeared.

> So when you look to which planet the first hour of the daytime belongs, then this planet is the ruler of that daytime. And when you look in the nights for the night whose first hour belongs to this same planet which ruled the first hour of the daytime, then this night belongs to this daytime.
>
> [III.203.26]

Now that we have understood the meaning of the 'taken-out' days, let us see which daytime was taken out of which night. As we pointed out above, Ibn 'Arabî argues that there are three other daytimes and three other nights between the daytime and the night from which it was taken out. That is because – as he explains – the structure of the world is six-directional: three nights corresponding to the directions down, left and back; and three daytimes corresponding to up, right and front [*Ayyâm Al-Sha'n*: 7]. Table 4.1 shows the resulting daytimes of the week and the nights from which those respective daytimes were taken out.

Therefore, as we can see, there are three days between each daytime and its partner night. Of course normally the observable earthly week will run starting from the beginning of Sunday night, then Sunday daytime, then Monday night, then Monday daytime, and so on. These are the normal 'circulated' days. The 'taken-out' days, however, as we see from this table above and in Figure 4.1, run also starting from the beginning of Sunday night but leaving out three daytimes and three nights jumping to Wednesday daytime (of the circulated days) and so on.

Therefore the flow of time for the taken-out days is different from the normal circulated days. In order to understand the flow of the taken-out days let us show the above relation graphically as in Figure 4.1. It is better to imagine this graph in three dimensions because, as Ibn 'Arabî indicated, the reason behind this interference between the taken-out days and the circulated days is the three-dimensional structure of the world that we live in [*Ayyâm Al-Sha'n:* 7].

But what is the significance of conceiving the 'taking-out' daytimes and nights in this way? In his book *Ayyâm Al-Sha'n* Ibn 'Arabî explains that the reason why there are three daytimes and three nights in the taken-out days is the three-dimensional (or six-directional) structure of the space in which we exist. In other books he (or rather, his later interpreter al-Qâshânî)[4] points out that the

Table 4.1 The taken-out days

	The daytime of . . .	*was taken out of the night of . . .*	
4	Wednesday	Sunday	1
5	Thursday	Monday	2
6	Friday	Tuesday	3
7	Saturday	Wednesday	4
1	Sunday	Thursday	5
2	Monday	Friday	6
3	Tuesday	Saturday	7

Note
There are three daytimes and three nights between the daytime and the night that is taken out of it. This table is summarized from *Ayyâm Al-Sha'n*, pages 6–7

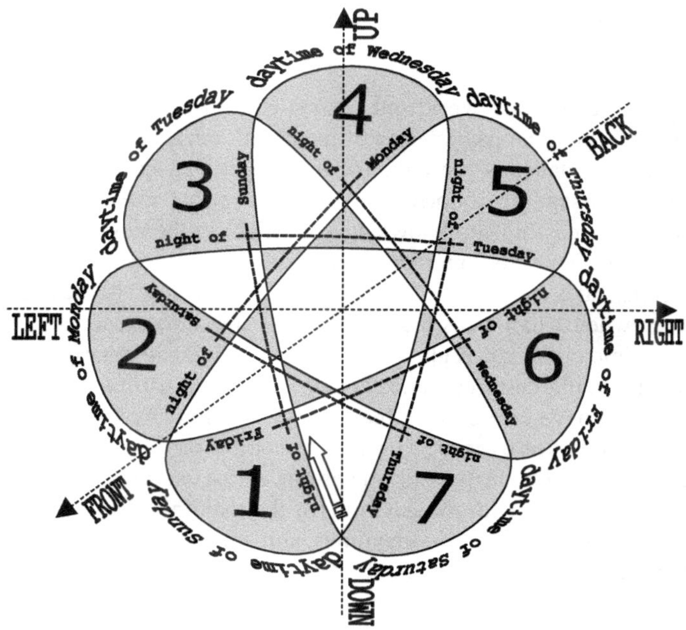

Figure 4.1 The taken-out days.

Note
It helps to imagine the resulting loop in three dimensions (x, y, z) as indicated graphically below.

word 'day' also means direction. So it appears to us as if our three-dimensional world is built up as a unit in seven Days (a Week). Therefore the Week with its seven Days is the unit of space-time, and not only time.

In chapter 302 of the *Futûhât*, Ibn 'Arabî explains how the process of taking the daytime out of the night is identical to taking the world out of non-existence into the 'light' of existence:

> Now we have mentioned that the world was hidden in the (absolute) Unseen (*ghayb*) of Allah, and that this Unseen was like the shadow of a person. So if something were taken out from the entirety of that shadow, it would come out in the image of that shade – just as that shadow is itself in the image of that of which it is a shadow. So the result of what is taken out of that shadow is in the image of the person. Do not you see that light is what appears when the daytime is taken out of the night-time?
>
> So those things which were hidden in the night (of the divine 'Unseen') appear by the light of the daytime. Therefore the daytime does not resemble the night, but rather resembles the light, by the appearance of those things through it. So the night was the shadow of the (divine existentiating) Light, and the daytime, when it is taken out of the night, comes out on the image

> of the Light. Likewise, the world, in its coming out of the Unseen, comes out in the image of the world (already present in) the Unseen, as we said. [III.12.1]

So the process of creating the world out of non-existence is exactly like the Qur'anic image of 'taking the daytime out of the night', because time in the end is reduced to the instantaneous presence (or the events that happen in it). As summarized in the passage just quoted, Ibn 'Arabî says that the world already exists in the all-encompassing foreknowledge of Allah, in the absolute Unseen (*al-ghayb al-mutlaq*). Thus its emergence into manifest existence is like the daytime: both of its states are 'Light', first hidden in the 'shadow' of the absolute Unseen, and then when Allah takes the world out of that non-existence, just as He 'takes the daytime out of the night'. And because Allah had determined that the manifest world would be three-dimensional, He created it 'in six Days' (or directions). He could have created it in many ways, but Ibn 'Arabî affirms that the divine Ability does not overrule His Determination or destiny (*qadar*); it only accomplishes what He has already determined (the *maqdûr*) [*Ayyâm Al-Sha'n*: 6]. Therefore, in order for the daytime (of the manifest world) to appear in existence, there have to be three daytimes and three nights between this daytime and the night that it was 'taken out' of [I.716; *Ayyâm Al-Sha'n:* 7]. Thus in each Day Allah creates a direction, and in the six Days of creation (from Sunday to Friday) Allah creates the world in (three-dimensional) space, while Saturday accounts for the 'Day of event' or time wherein this world is displayed.

4.4 The 'intertwined' days

As we have mentioned before (section 2.15), Allah said in the Holy Qur'an: '*each Day He is upon some (one) task*' (55:29). Since Allah did not say 'task*s*' but rather a single task or event, Ibn 'Arabî argues that the whole Day should be under the effect of one single divine task. This, however, is not the case for our normal, 'circulated' days and is clearly not the case for the 'taken-out' days we have just described above, because we know that many different events are happening in each observable 'circulated' day.

Ibn 'Arabî argues that the original Days of events in which Allah described Himself as being '*each Day upon some task*' are *intertwined* (or 'entered into each other': *mutawâlija*, from the verb *yûliju*: 'to enter into') with the circulated days in a specific manner, which he explains by dividing the day into 24 hours. However, Ibn 'Arabî emphasizes that this example is only for approximation, since one could also explain this 'intertwining' on a smaller scale than hours (e.g. minutes and seconds, or even smaller). The matter as he describes it is already very complicated for 24 hours, although this may be possible to calculate now using sophisticated computer programs.

Starting with the night of Sunday – because its name *al-ahad* ('the First', 'the Unique') is a Name of Allah, it is the first day, and also it is the day of the Sun that is the heart or centre of the manifest world [*Ayyâm Al-Sha'n*: 11] – Ibn

'Arabî reconstructs the Days of events from the hours of the circulated days, starting with the first hour of the night of Thursday, then the eighth hour, and so on with seven-hour intervals until the full 12 hours of the night of Sunday are completed. Then he moves on to 'deconstruct' the daytime of Sunday in the same way, as illustrated in the Table 4.2 and Figure 4.2.

Then he moves on to analyse the night of (the circulated day of) Monday in the same way, but starting from the daytime of Friday, and so on for the full seven Days, as indicated in the following Table 4.3 and Figure 4.2.

4.4.1 Demonstrating the intertwined days

As can be readily seen from Tables 4.2 and 4.3, the flow of time for the 'intertwined' days is even more complicated than that of the taken-out days. This

Table 4.2 The intertwined days (example of Sunday alone).

	Normal week days						
	SUN	*MON*	*TUE*	*WED*	*THU*	*FRI*	*SAT*
Hours of night							
1	–	–	–	–	**SUN**	–	–
2	–	–	–	–	–	–	**SUN**
3	–	SUN	–	–	–	–	–
4	–	–	–	SUN	–	–	–
5	–	–	–	–	–	**SUN**	–
6	**SUN**	–	–	–	–	–	–
7	–	–	SUN	–	–	–	–
8	–	–	–	–	**SUN**	–	–
9	–	–	–	–	–	–	**SUN**
10	–	SUN	–	–	–	–	–
11	–	–	–	SUN	–	–	–
12	–	–	–	–	–	**SUN**	–
Hours of daytime							
1	SUN	–	–	–	–	–	–
2	–	–	SUN	–	–	–	–
3	–	–	–	–	**SUN**	–	–
4	–	–	–	–	–	–	**SUN**
5	–	SUN	–	–	–	–	–
6	–	–	–	SUN	–	–	–
7	–	–	–	–	–	**SUN**	–
8	SUN	–	–	–	–	–	–
9	–	–	SUN	–	–	–	–
10	–	–	–	–	**SUN**	–	–
11	–	–	–	–	–	–	**SUN**
12	–	SUN	–	–	–	–	–

Note

The shaded background indicates the nights of the normal days, and bold font indicates the night-time hours of the intertwined days. The data in this table are extracted from *Kitâb Ayyâm Al-Sha'n*, pp. 11–12.

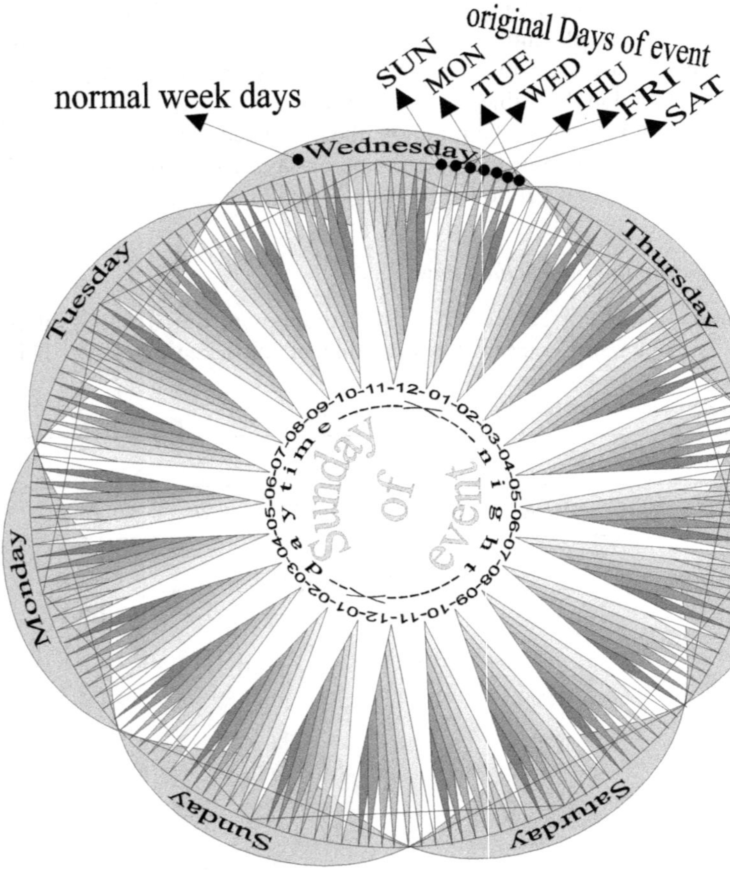

Figure 4.2 The intertwined days, and their relation with the circulated days.

Note
The information in this figure is extracted from *Kitâb Ayyâm Al-Sha'n*, pp. 11–16.

complex relationship is shown graphically in Figure 4.2. Again, it would be more accurate to imagine this graph in three dimensions, but the graphic depiction is already very complicated in two dimensions. (A more accurate approach to what Ibn 'Arabî is actually describing would be to simulate this on the computer, taking into account smaller time-scales than hours.)

Figure 4.2 demonstrates how the 24 hours of each day of the normal circulated days are distributed over the seven Days of events. The 24 hours of the Day of Sunday for example (the 24-fold zigzag circle in the front) are distributed over the normal week days, starting from the first hour of Thursday night and moving with the seven-hour intervals already described.

Table 4.3 The intertwined days (all days)

	Normal week days						
	SUN	*MON*	*TUE*	*WED*	*THU*	*FRI*	*SAT*
Hours of nights							
1	**WED**	**THU**	**FRI**	**SAT**	**SUN**	**MON**	**TUE**
2	**MON**	**TUE**	**WED**	**THU**	**FRI**	**SAT**	**SUN**
3	SAT	SUN	MON	TUE	WED	THU	FRI
4	THU	FRI	SAT	SUN	MON	TUE	WED
5	**TUE**	**WED**	**THU**	**FRI**	**SAT**	**SUN**	**MON**
6	**SUN**	**MON**	**TUE**	**WED**	**THU**	**FRI**	**SAT**
7	FRI	SAT	SUN	MON	TUE	WED	THU
8	**WED**	**THU**	**FRI**	**SAT**	**SUN**	**MON**	**TUE**
9	**MON**	**TUE**	**WED**	**THU**	**FRI**	**SAT**	**SUN**
10	SAT	SUN	MON	TUE	WED	THU	FRI
11	THU	FRI	SAT	SUN	MON	TUE	WED
12	**TUE**	**WED**	**THU**	**FRI**	**SAT**	**SUN**	**MON**
Hours of daytimes							
1	SUN	MON	TUE	WED	THU	FRI	SAT
2	FRI	SAT	SUN	MON	TUE	WED	THU
3	**WED**	**THU**	**FRI**	**SAT**	**SUN**	**MON**	**TUE**
4	**MON**	**TUE**	**WED**	**THU**	**FRI**	**SAT**	**SUN**
5	SAT	SUN	MON	TUE	WED	THU	FRI
6	THU	FRI	SAT	SUN	MON	TUE	WED
7	**TUE**	**WED**	**THU**	**FRI**	**SAT**	**SUN**	**MON**
8	SUN	MON	TUE	WED	THU	FRI	SAT
9	FRI	SAT	SUN	MON	TUE	WED	THU
10	**WED**	**THU**	**FRI**	**SAT**	**SUN**	**MON**	**TUE**
11	**MON**	**TUE**	**WED**	**THU**	**FRI**	**SAT**	**SUN**
12	SAT	SUN	MON	TUE	WED	THU	FRI

Note
The shaded background indicates the nights of the normal days, and bold font indicates night hours of the intertwined days. The data in this table are extracted from *Kitâb Ayyâm Al-Sha'n*, pp. 11–16

4.4.2 The significance of the intertwined days

The flow of time according to the intertwined days is the real flow, because it indicates the manifest, outward order of events. The reason why there is a constant seven-hour step, as in the Tables 4.2 and 4.3 and Figure 4.3, is that the order of creation proceeds according to the seven main divine Names (and divine 'Days' of creation). For example, on Sunday (of the original Days of creation) the world starts 'hearing' the divine Command, which initiates the first dimension in creation. Then once the Sunday cycle (i.e. of the divine Attribute 'the All-Hearing') is over, the Monday cycle starts granting the world the attribute of 'the Living', and so on (see Table 3.1). Thus the manifest world gains the qualities of these fundamental divine Attributes one after the other.

4.5 The motion of the seven days in the zodiac

The orb of the 'fixed stars', including the constellations of the zodiac and the lunar mansions, is the second orb after the outermost Isotropic Orb, but it is the first material orb, because the Isotropic Orb does not have any distinguishing features (no stars or planets). We have already noted that Ibn 'Arabî shows that 'Allah created the Day in the first orb, and then defined it in the second orb, which is the orb of the fixed stars (the zodiac)' [*Ayyâm Al-Sha'n:* 6]. That is because the motion of the Isotropic Orb can not be defined for the world beneath it, because there is no reference point to assign a beginning or an end to this motion. However, in the zodiac orb there is such a reference point: the 12 Signs or constellations. So we can say that the days were first defined in the zodiac orb, and each day represents a full cycle of this orb as observed from the Earth [III.548.31; see also section 2.12 above]. Ibn 'Arabî clearly showed that this orb does not actually move as we observe it [II.441.33, I.141.17, III.549.3], but that we see it as moving because of the motion of the Earth around its centre [I.123.17, II.677.21, III.548.31] (see also section 1.4).

Ibn 'Arabî takes the first degree in Gemini as the reference point to start counting the cosmic 'Days' of creation, and he claims that this initial motion of time in creation started when this point was matching the divine 'Foot' in the 'Pedestal' above the Isotropic Orb. It is very interesting to note that the first degree of Gemini coincides with the galactic equator.[5] Moreover, recent studies and some historical or archaeological records (Sitchin 1976: 197) show that this point was used by ancient Sumerians as early as 11000 BC as a starting point for the zodiacal calendar. He says:

> And let it be known to you that this Name ('the All-Sufficient', *al-ghanî*) made this orb isotropic with no planets in it: all its parts are the same, and it has a circular shape where you can not distinguish a beginning or an end to its motion, and it has no edge. By the existence of this orb the seven days came into being, and so months and years; but these (particular durations of) times were not determined in it until Allah created, inside this orb, other signs (galaxies, stars and planets) which determined these times.
>
> This orb determined only one Day, which is a single cycle (of this orb) beginning at the place of the (divine) 'Foot' in the Pedestal. So it was determined from above. The duration of this cycle is called a Day. But nobody except Allah knows (the actual length of) this Day, because of the fact that all the parts of this orb are the same, so (nobody knows) the beginning of its motion. And the beginning of its motion was when the first degree of Gemini, which is an orb (whose influence is associated with the lower material element) of air, was matching this Foot.
>
> Hence the first Day that appeared in the world was by the first degree of Gemini, and this Day is called *al-ahad* (Sunday). So when this specific point of this orb, which is known only to Allah, returned to match this Foot of the Pedestal, a full cycle – that is, of the total orb (and its contents) – was com-

> pleted and all the parts of this (isotropic) orb matched with the (initial) place of the Foot from the Pedestal. So this motion came over every degree, minute, second, and less than that of this orb. So the places came into being and the existence of the single indivisible localized monad (of all manifest creation) was determined from the motion of this orb.
>
> [II.437.29]

That was the motion of the initial Day of the Sun (Sunday).

> Then, at the end (of this initial Day), there began another motion, also from the middle,[6] until that (second motion) reached its end/aim (*ghâya*), like the first motion – together with the entirety of what (that sphere) contains of the (smallest) parts (*ajzâ'*) and the individual entities that are composed of those parts, because (that sphere) is quantitative. This second motion is called Monday. And so on till it completed seven periodical motions everyone is determined by a divine Attribute. And the (main) divine Attributes are seven, not more, which made the Age not more than seven (distinctive) Days.
>
> [II.438.3]

This is shown better in Figure 4.3.

Thus the orb of the zodiac (fixed starts) executes only seven distinctive cycles, which are then repeated that over and over again. Each Day of these seven Days of the divine creative Week is called a 'Day of event', in which only one single event happens, because it is ruled by one primary divine Name of Allah. So there are seven Days of events, each related to one particular divine Name from the seven fundamental Names mentioned in section 3.4.

4.6 The kinds of events

Now in each 'Day of event' that is intertwined with the outwardly observable (normal earthly) 'circulated days' as described above, Ibn 'Arabî explains that Allah constantly inspires the Universal Soul to act upon the orbs below it – encompassing and giving rise to all the manifest, material cosmos – so that they move in a specific manner that will cause one particular kind of event in the cosmos. But the *effects* of this complex composite motion of the orbs will be different, depending on the capabilities and characteristics of individual creatures [*Ayyâm Al-Sha'n*: 11–12]. For example, when Allah inspires the Soul to move the element of fire in order to heat the world, the effects of this single event depends greatly on the individual creatures: those which are ready to burn will burn, and those which accept heat will be heated, and so on [*Ayyâm Al-Sha'n*: 6], and he explains further in the *Futûhât* that:

> the (divine creative) "Event", in relation to the Real (i.e., God/*al-haqq*), is one from Him; but with relation to the acceptors of the world it is many events that we would call infinite, were they not confined by (their shared quality of) existence.
>
> [II.82.6]

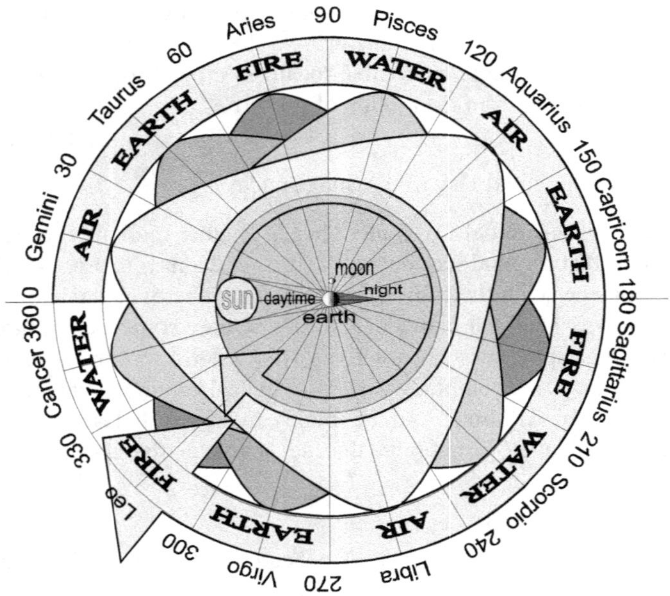

Figure 4.3 The Zodiac and the motion of the days in it.

Note
Notice that this refers to the apparent rotation of the sphere of the fixed stars, while the actual astronomical motion is due to the revolution of the Earth. The 12 zodiacal signs are divided into four groups whose influences are related to the four material elements (earth, water, air and fire), as described in the preceding chapter; each group contains three signs of the same nature. The motion of each cosmic Day starts at the first degree of Gemini.

Ibn ‘Arabî describes in his *Kitâb Ayyâm Al-Sha’n* the different kinds of events that are particularly associated with each Day of the seven Days of the original divine ‘Week’ of event. There he also gives a ratio of ‘contribution’ by each one of the seven heavens. These mysterious symbolic data and concepts are not very easy to understand. However, we will summarize his remarks in Table 4.4 for reference, although any attempt to interpret them would require a separate extensive study.

The data in Table 4.4 are extracted from *Kitâb Ayyâm Al-Sha’n*: 11–16. The contributions are stated to be increments of one-fourth, probably alluding to the equal role of the four basic elements of material Nature (fire, air, water and earth). The −/− sign means that the corresponding data are not given in the text. For Wednesday, Ibn ‘Arabî does not give detailed numbers, but simply says (p. 14) that ‘God ordered the spiritual realities to help the Soul according to their corresponding strength’. The ‘↓’ indicates that the contribution is done only by way of descent. Some of the data in this table are incomplete or not certain, as the meaning of the original text is often unclear.[7]

Table 4.4 The action of the soul (on the world) in each day of the seven days of events and the ratio of contribution by the seven heavens

	SUN	*MON*	*TUE*	*WED*	*THU*	*FRI*	*SAT*
The actions of the Universal Soul in each Day of event							
	moving the fire element to heat the world	giving the element of liquids to the generators	showing the absurd passions	mixing wet vapour with dry vapour	liquefying and decomposing	distilling the wet vapour	holding the forms of the world and forming them
The ratio of contribution by the seven celestial spheres (orbs)s							
7	2/4	2/4↓	2/4	–/–	0/4	2/4	4/4
6	2/4	2/4↓	2/4	–/–	4/4	2/4	0/4
5	4/4	0/4	–/–	–/–	2/4	0/4	2/4
4	4/4	0/4	4/4	–/–	2/4	0/4	2/4
3	0/4	4/4	0/4	–/–	2/4	4/4	2/4
2	1/4	1/4+1/4	–/–	–/–	2/4	2/4	2/4
1	0/4	4/4	0/4	–/–	2/4	4/4	2/4
The Results							
	the spirits flow in spiritual beings, and motion in the things that move	bodies grow; rainy winds blow; motions are weak	anger and conflicts amongst people	–/–	good for lovers	the events result through the divine Command, not as the result of trying and direct action	–/–

4.7 Hours, minutes and seconds

According to what we have now discussed in this and preceding chapters, the day as we normally experience or observe it is the minimum period of time defined by the apparent full revolution of the Heavens around the Earth (see sections 2.12 and 3.2). The original divine creative Days are single indivisible divine 'Acts', but in the observed time that we experience, we live only one moment out of each original Day (see section 2.15). Therefore, the normal day as we know it is a collection of the moments from the different divine Days of the creative Week, after the processes of 'intertwining' and 'taking-out' detailed above. For this reason, the normal day can be conventionally subdivided apparently infinitely, and smaller units of time can be conceived and measured.

Thus, the day is conventionally divided into 24 hours, and the hour is conventionally divided into 60 minutes, which are also conventionally divided into 60 seconds each. This conventional hexadecimal system can be traced back to the time of the Babylonians; it was used by many ancient civilizations and then transferred by the Arabs to the West. In recent centuries, some more scientifically measurable definitions of these conventional units of time have been introduced, specifically for the second.[8] So this hexadecimal system is now internationally used, but after the second, smaller units of time are decimally subdivided into 0.1, 0.01 etc., or 10^{-x} seconds, where x could be in principle any number no matter how large (for infinitely shorter times).[9]

Ibn 'Arabî referred to these further conventional subdivisions of observed time in chapter 59 of the *Futûhât*, which is entitled 'On the Inner Knowing of the (actually) Existent and Conventional Time'. Here the original divine cosmic 'Days of event' are the actually Existent Time (*al-zamân al-mawjûd*) while the other time-distinctions days, years, months, hours, minutes and seconds are only conventional or 'estimated' (*muqaddar*). There he explains:

> The (observable) 'days' are many: some are long, and some are short. The shortest day is the 'monad of time' (*al-zaman al-fard*) in which '*He is in a task*' (55:29), so that (indivisible shortest) monad of time is called a 'day' because the (divine) 'task' happens in it: it is the shortest and tiniest time, and there is no maximum term to the longest. Between them there are intermediate 'days', the first of which is the conventional 'day' familiar in customary usage, which is divided into hours, the hours into degrees, the degrees into minutes, and so on. That goes on indefinitely, for those people who also divide the minutes into seconds. Since it is ruled by (the principle of) countability, (for them time) is like numbers; and numbers are infinite, so this division is also infinite.
>
> But some people say that (time) is finite, and they take this matter from the perspective of what can (actually) be counted or 'numbered': those are (the people) who advocate that time is an existing essence (because it is countable), and all that exists must undoubtedly be finite. But the opposing (group) say that the countable, simply by the fact that it can be counted, does not

(therefore necessarily) enter into existence, so it is not described as finite, because number is not describable as being finite. This is how those who argue against the 'single monad' (or 'indivisible atom': *al-jawhar al-fard* proposed by the kalam theologians) contend, (maintaining instead) that according to the intellect body is divisible indefinitely. This topic of controversy among the people of intellectual inquiry (i.e. kalam theologians and the philosophers) happened as a result of their lack of good judgement and their inquiring (only) about the meanings of words (instead of what is actually real).

[I.292.3]

However, for Ibn 'Arabî this conventional division of time also has its divine origin. He first affirms that the celestial orbs were divided into 360 degrees due to the fact that the Universal Intellect (or the 'Higher Pen') was taught by the 'Greatest Element' (see sections 6.2 and 6.5) exactly 360 kinds of comprehensive divisions of knowledge (*'ulûm al-ijmâl*), under each of which there are 360 further divisions of detailed knowledge (*'ulûm al-tafsîl*), that is 360 times 360:

So each degree (of the 360 total degrees of the outermost, all-encompassing celestial sphere) includes all that it comprises of the details of minutes, seconds, tertiary divisions, and so on, as Allah – the Exalted – wills to make manifest (of that detailed Knowledge) in His Creation till the day of the Rising.

[I.295.12]

After that, in the same passage, Ibn 'Arabî shows that Allah appointed 12 'rulers' (*wulât*, s. *wâli*), one in each of the 12 zodiac signs, which correspond to the 12 months in the conventional year. Also there are 28 'chamberlains' (*hujjâb*, s. *hâjib*) in the orb of the constellations (i.e. the 28 houses of the Moon) which correspond to the 28 days in the (divine) month (see section 3.2). Finally, in the same passage he also says that Allah ordered those rulers to appoint seven 'chiefs' (*nuqabâ'*, s. *naqîb*) in the seven Heavens, one 'chief' in each celestial sphere – i.e. the seven orbs which correspond to the seven Days of the cosmic Week. Ibn 'Arabî then adds that each one of those groups of spirits that are associated with the celestial spheres has a living human 'deputy' or 'agent' among us on Earth [I.296.27].

Ibn 'Arabî even suggests a certain divine importance for the conventional division of the day into 24 hours, in the following passage discussing the different groups of the spiritual hierarchy of the 'Friends of God' (*awliyâ'*) in chapter 73 of the *Futûhât* [II.2.39]:

And among them (the *awliyâ'*), may Allah be pleased with them, (are a group of) 24 souls at any time, no more and no less. They are called the 'Men of (Spiritual) Opening' (*rijâl al-fateh*): through them Allah opens for the people of Allah whatever He opens of (divine) knowledge and secrets. And Allah made them according to the number of hours, one of them for each hour.

[II.13.7]

Regarding the hours of the day and the two parts of it (the daytime and the night), and their observable variations across the year and from one place to another, Ibn ʻArabî says:

> Then Allah caused the (observable, earthly) daytime (*nahâr*) and the night-time by the existence of the Sun – not the (true divine) Days (*ayyâm*). And as for what happens of (the observable) increase and decrease in the (length of the) daytime and night – not in the hours, for they are (always) 24 hours – that is because of the motion of the Sun in the zodiac region which is tilted with regard to us, so the day becomes longer where it is in the high houses, and when it comes to the low houses the day becomes shorter where it (i.e. the Sun) is. We said 'where it is', because when the night becomes longer for us the day becomes longer for others, so the Sun is in the high houses for them and in the low houses for us. So when the day becomes shorter for us, the night becomes longer for them as we said. But the day (*al-yawm*) itself is 24 hours: it does not increase or decrease, nor does it become longer nor shorter in the equinox place.
>
> [I.140.33]

5 Unicity and multiplicity

Examine the origin of existence, and think it through:
you will see the same (divine) Generosity, both eternal and created.
And the (created) thing is like the (eternal) thing,
except that He made it appear in the actual reality of the worlds as created.
So if the viewer swears that his existence is eternal (in the Unity),
he is honest and truthful, not lying;
But if the viewer swears that its existence (emerges) after its non-existence,
(that is) more appropriate – and then its [existence] *is threefold.*

[I.5.31]

In order to understand these novel concepts of time that we have explained in previous chapters, we need to shed more light on Ibn 'Arabî's controversial understanding of the oneness of being (*wahdat al-wujûd*), because it is the key to understanding his various views of time. Although he had never employed this famous term directly, it is quite evident that this characteristic understanding of the oneness of being strongly dominates Ibn Arabî's many writings: he explains almost everything on the basis of the concept of ultimate unicity and oneness. More specifically, the metaphysical structure of the world, how it comes into existence, how it is maintained and its ontological relation with the Creator can be explained only on the basis of the oneness of being.

In the light of the centuries of later polemics surrounding this conception, it must be stressed that Ibn 'Arabî's conception of the oneness of being is completely different from such views such as pantheism or monism. Ibn 'Arabî never denies the basic distinction between the Creator and creations, and certainly never contradicts the basic tenent of monotheism. Likewise, he strictly affirms the Muhammadan revelation (*sharî'a*) and accepts all the observed multiplicity on the usual planes of manifest existence. Instead, his focus in applying the oneness of being is on understanding the cosmos and how it works. It is these particular cosmological aspects that we shall be interested in throughout this chapter, so that we can explain Ibn 'Arabî's view of the creation in order to understand his unique concepts of time and eventually to situate all those conceptions within the comprehensive cosmological model outlined in Chapter 6 below.

5.1 The contrasting approaches of Sufism and philosophy

We want to start this chapter by quoting Ibn 'Arabî's famous story of his first encounter with the great Aristotelian philosopher Ibn Rushd (Averroes, already mentioned in section 1.7) when Ibn 'Arabî was still relatively young, but already famous for his immense knowledge and unique views. According to that account, Ibn Rushd arranged with his friend – Ibn 'Arabî's father – to meet the young mystic in order to hear what he had to say about (his Aristotelian) philosophy. As Ibn 'Arabî recounts this story in the *Futûhât*:

> And I entered one day in Cordoba into (the house of) the judge Abû al-Walîd Ibn Rushd, as he wished to meet me after he had heard about what Allah opened up for me in my spiritual retreat, for he had been expressing his admiration of (or 'amazement at') what he had heard. Then my father – because he was one of (Averroes') friends – sent me to him for something on purpose in order (for him) to meet me. I was still young; my face had not yet put forth a beard, and my moustache had not yet grown.[1]
>
> When I entered to see him he stood up for me out of love and respect, embraced me and said (exclaiming): 'Yes'! I replied: 'Yes'. So his joy was magnified because I understood him. Then I realized what made him feel happy, and I said: 'For Allah's sake, No!' Then he turned sad, his colour changed and he doubted his philosophy.
>
> Then he asked: 'So how did you find it in unveiling (*kashf*) and divine effusion (*fayd ilâhî*)? Is it the same as what thought had led us (philosophers) to?' I replied: 'Yes … No, and between the "yes" and the "no", spirits fly away from their (bodily) matter and necks from their bodies.'
>
> So his (face) colour turned pale, he began to tremble and sat down reciting the *hawqala* [that is to say: *lâ hawla wa lâ quwwata illâ bi Allah* ('there is no power and no strength but in Allah')], and he knew what I alluded to (in responding) to him.
>
> [I.153.33]

This mysterious exchange of these few words and gestures between these two pillars of Islamic thought, a Sufi and a philosopher, is an attempt to express in symbolic language what is very difficult to explain in more explicit language. Ibn 'Arabî alludes here to an essential realization that is beyond normal human comprehension, something that is apparently against our everyday experience or otherwise very difficult to believe. Yet on the other hand, it is something that can be ultimately summarized in only two words: 'Yes' and 'No', or even 'Yes' alone, because 'No' is 'not Yes'.[2] In fact, Ibn 'Arabî's 'digital' answer here: 'Yes/No' (or '1/0', 'True/False', which ultimately amounts to existence/non-existence) is the best and shortest expression of the creation – summarizing the essence of the paradoxical metaphysical insights alluded to in the initial poem of the *Futûhât* translated at the beginning of this chapter.

The difficulty of expressing this universal Reality in simple words comes from the fact that we live in a diverse world of infinite multiplicity, while at the same time the truth or reality behind this world is literally too simple to be believed [*Al-Masâ'il*: 163]. The ultimate Real is Allah, and Allah is uniquely One, while the world is apparently many – so the metaphysical challenge is how to link the (imaginary) multiplicity of the world to the Real One, through some unseen intermediaries.

Philosophers and scientists in general try to understand the world through observations, while the methods of Ibn 'Arabî and other Sufis rely upon modes of perception that jump directly into the unseen in order to approach the Real directly. As Ibn 'Arabî often points out, observations are subject to many mistakes, owing to the inaccuracy of the tools employed, whether human senses or technical equipment, while *true* visions – as opposed to our sometimes problematic interpretations of them – are always correct [I.307.12, III.7.21].[3] On the other hand, philosophers and scientists use logic and experiments to deduce their theories and explain their observations, while Sufis in general often describe their visions without paying too much attention to explaining them in a logical manner, especially when some of their visions, though real and true, may be outwardly or apparently illogical.[4] As a result, certain Sufis like Ibn 'Arabî may attain a very high state of knowledge of reality more quickly and more accurately than philosophers (as Ibn 'Arabî certainly implies in his account of his meeting with Ibn Rushd), but they find it very difficult to explain their views to others who have not '*tasted*' it their way. So when they try to explain their insights, not many people will understand what they say.

The problem with the current laws and theories of physics and cosmology is that so far, although they have proved to be quite accurate and powerful in application, they have admittedly failed to unveil the ultimate reality behind the world. All scientific theories are descriptive rather than determinative. We have seen in section 1.3 that the reason why science was not able to determine the reality of the world is that all cosmological models need a boundary condition: i.e. an exact description of what was the initial state when the world started, something which seems to be impossible to achieve by the intellect alone. That is why scientists work backwards: i.e. they try to find out the initial state of the cosmos by extrapolating in various ways from the current observations. As a result, all physics theories and known cosmological models, though they have achieved higher levels of understanding, have also brought new contradictions and paradoxes. They have succeeded in providing approximate possible creation scenarios, but failed to describe the reality itself. Although Ibn 'Arabî considers the intellect unbounded or unlimited as a *receptive* tool, it is quite limited as a ratiocinative thinking tool because it relies on limited senses [I.288.27]. Therefore, the intellect alone – as a thinking tool – cannot describe the origin of the world because it is necessarily a part of it. Ibn 'Arabî affirmed this when he said that the limit of the observations of the philosophers (or astronomers) is up to the Isotropic Orb which is the first (outermost, and first created) material orb [II.677.1]; they can not see or detect anything beyond that. That is why the Sufis rely on the 'heart' (the locus of spiritual 'tasting' and inspiration, in the language of the Qur'an) rather than the discursive intellect.

On the other hand, Sufis have sometimes claimed to have achieved a high state of realization and even to have visualized the metaphysical structure and origin of the world (i.e. what physicists call the initial/boundary condition). Most of them, however, did not give proper attention to explaining the observable universe and relating it to that initial cosmological state. Even Ibn 'Arabî himself did not care too much about that: instead he declared that his aims were not to explain the world, but rather to acquire more knowledge of the world as a structure created according to the Image of Allah, so that he might acquire more knowledge of Allah Himself. All the same, however, throughout the *Futûhât* and other shorter books Ibn 'Arabî often gives a great many cosmological explanations and sometimes logical analyses of his metaphysical visions. This is why it is very important to study Ibn 'Arabî's writings, since they may provide a real link between philosophy and science, on the one hand, and mysticism and theology.

5.2 Unicity versus multiplicity

Ibn 'Arabî cited the story of his meeting with Ibn Rushd in the context of explaining the words of the central spiritual Pole Idrîs (*mudâwi al-kulûm*) who – as Ibn 'Arabî said – knows very well about the natural world and the effects of the higher world on it. Thus this Pole explained that 'the world exists between the circumference and the point' [I.154.22]. The 'point' here refers to the Real (the 'Necessary Being') whose existence is self-existence, while the 'circumference' is the circle of creations (the 'possible' or contingent entities) whose existence depends on the Real. Beyond this circumference is the 'sea' of non-existence (the impossible of existence). This relationship is illustrated in Figure 5.1.

With regard to this image, Ibn 'Arabî explained in chapter 47 of the *Futûhât* [I.260.1] that the (divine creative Source-) point in the centre of a circle meets any point in its circumference with its whole entity, without division or multiplicity. Similarly, multiplicity (i.e. all of creation) appears or emerges out of the Unicity of the Real; the manyness of the world appears out of the One Creator, without affecting His unique Oneness or Unicity. Ibn 'Arabî was well aware that this paradoxical relation between the Creator and all manifestation is in clear apparent contradiction with the widely accepted philosophical maxim – a central assumption in the prevailing contemporary philosophical cosmology of Ibn Sina and his followers – that 'from the One only one may emerge (or proceed)' (*la yasdur 'an al-wâhid illâ wâhid*).[5]

Given the assumption of this maxim, an obvious problem encountered by philosophers and theologians when they want to explain how Allah created the world is that Allah is One while the world is many. So logically it is not possible to imagine a relation between the One and the many without affecting the unique Oneness (*ahadiyya*) of the One.

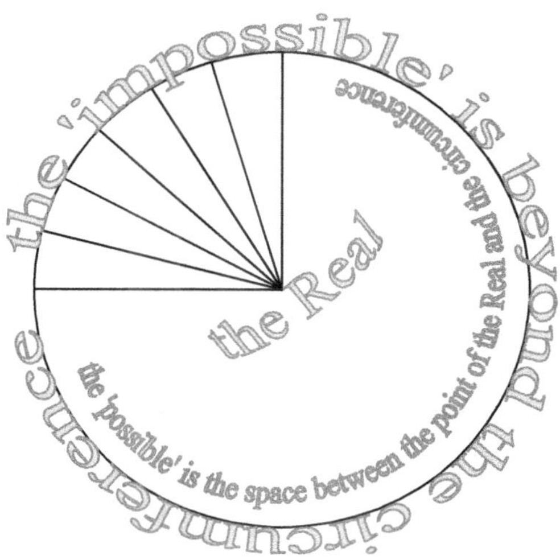

Figure 5.1 The Real, the 'Possible' existents, and the 'Impossible'

Note
This figure is taken from chapter 360 of the *Futûhât* [III.275].

However, Ibn 'Arabî's analogy between the pure mathematical symbol of the circle and its centre and the cosmological process of creation by the One Creator is not fully justifiable without further explanations. Among other problems, mathematics and geometry work with infinitely small (or dimensionless) points, while our contemporary science of physics and cosmology deals with corporeal worlds that have dimensions. But we shall see below that Ibn 'Arabî's unique understanding of time provides here the essential link between physics and mathematics, or between reality and imagination, in the same way as it does provide the necessary link between unicity and multiplicity.

Ibn 'Arabî quotes the above-mentioned emanationist philosophical maxim quite often [I.42.14, I.260.5, II.31.14]. Although he disagrees with this general proposition [I.260.5, I.715.12, II.434.20, and see also *Al-Durratu Al-Baydâ'*: 140], he sometimes explains further that this notion can be held true for physical beings but not for Allah Himself, because Allah, the unique One, can obviously create multiple creations as we can clearly see:

> So without their dependence (for actual existence) on the existing-entity (*'ayn*) of the servant, there would be no rulership for those two Names ('the First' and 'the Last'). Because there (in eternity), the (divine) Essential-entity (*al-'ayn*) is (uniquely) One, not united (from different parts: *muttahida*). But in the servant, (the existing-entity) is united (of different parts) and not (uniquely) one, because Oneness (*al-ahadiyya*) is for Allah (alone), and unification (*ittihâd*, the unification of the servant's parts, senses, faculties etc., and not the unification of the servant with his Lord because this is

> not possible in Ibn ‘Arabî’s doctrine) – and not the (divine) Oneness – is for the servant. This is because the servant can only be understood in relation to another (Who is his ‘Lord’ or his Creator), and not by himself: so he has no trace of (the absolute divine) Oneness at all. But as for the Real, Oneness may be understood (as applying) to Him (taken by Himself), or it may be understood (as applying to Him) with relation (to others), since everything belongs to Him, and indeed He is actually the Essence of everything. (This unique divine Oneness refers) not to the wholeness uniting a collection (of different entities: *kulliyyat jam‘*), but rather to the (unique) Reality of Unicity (*haqîqat ahadiyya*) on which (all) multiplicity depends – and this (unique Oneness) can only apply specifically to the (divine) Real.
>
> So according to the determination of the (human) intellect, only one thing can ever emerge from the One. But the Unicity of the Real does not fall under that rule. How could He, Who created that rule, fall under it?! And the (true) Ruler – *there is no god but Him, the Almighty, the All-Wise* (3:6, 3:18).
>
> [II.31.11]

So now we can see how multiplicity may come out of the Oneness of the Real. Yet we need to explain how this multiplicity of the creation appears from the One Creator. We need to explain how the pure geometrical analogy of the circle and its centre could be applied to the creation of the worlds by Allah.

Ibn ‘Arabî solves this riddle in part by inserting ‘time’, a true understanding of time. Hence he says:

> and He (the Real) has a special face (*wajh khâss*) towards everything that exists, because He is the cause of everything. Now every (single) thing is one, it cannot be two; and He is One. So from Him there appeared only one, because He is in the oneness (*ahadiyya*) of every one (existing thing).
>
> So if multiplicity exists, it would (only) be with regard to the *oneness of time* that is the container (of that apparent multiplicity). For the existence of the Real in this multiplicity is in the oneness of every one (existent). So there appeared from Him only one. Therefore this is the real meaning of ‘from the One only one may emerge’: even if the entirety of the world appeared from Him, there would only appear from Him one (created reality), because He is ‘*with*’[6] every one (of the creatures) with respect to its oneness.
>
> Now this is something that can be perceived only by the (truly enlightened) ‘people of Allah’,[7] whereas the philosophers mean this [i.e. that from one (cause) only one (effect) can emerge] in an entirely different sense, and this is something about which they were mistaken.
>
> [II.434.18]

Because of the rarity of the underlying spiritual perception of this reality restricted, as Ibn ‘Arabî stresses, to the fully enlightened ‘people of Allah’, this passage just quoted above is not readily understood. Perhaps it is because of its great importance and central role in Ibn 'Arabî's cosmology that the famous 'Abd-al-Qâdir al-Jazâ'irî,[8] the editor of the original Bulaq edition of the *Futûhât* that is the basis for all modern versions, added a rare long footnote comment at this point in the *Futûhât* [II.434–435] to explain it in terms of the oneness of being. Because his comments here are very helpful in this regard, we shall analyse them at length in what follows.

Here he explains that this passage refers to two related issues: 'the one-ness of every being' (*wahdat kull mawjûd*) and 'the unicity of Being (*ahadiyyat al-wujûd*)'. He begins by pointing out that everyone and every thing has a unique 'face' or individual reality that makes it distinctive; thus there are no two persons or entities with the same reality, since otherwise they would be one (the same), and not two. So this means that there appears from the One Creator only one (reality) because this unique 'special face' is never repeated (see also section 2.8).

This explanation, however, is not entirely satisfactory (for our purposes, at least), because the negation of repetition does not imply the negation of multiplicity, which is clearly witnessed in the world.

'Abd al-Qâdir then goes on to explain that we can reconcile the apparent multiplicity with the actual oneness of creation by correcting our view of time and space. He says that our imagination pictures time as a container which contains the things in the existence (as Newton later imagined time, but the theory of Relativity superseded his view, as we mentioned in section 1.3) so we see things arranged in time (and space), and then we imagine multiplicity. But if we imagine ourselves 'out of time', and look at the whole existence in time and space, we shall see a single existence without a beginning and without an end, and without any relation to a self-subsistent distinct time and space as we usually imagine them. For example, every person is one despite having arms and legs and many visible and invisible parts.

But again this explanation is not entirely satisfactory, since it shows the unity of all being (*wâhidiyya*), but not its unique metaphysical 'unicity' (*ahadiyya*). It shows that the whole of existence is 'one' when we look at it as a single whole, or from outside space and time. But still, since we actually perceive (or imagine) ourselves as existing inside this space-time whole as partial entities, we also see many other entities – or in other words, manifest multiplicity. So we still need to explain how this multiplicity appears from the unique Oneness of the Real.

'Abd al-Qâdir then goes on to explain that if the philosophers meant by saying 'from the one there only emerges one' that Allah created only the First Intellect (which is the way this maxim was understood by Ibn Sina and most Islamic philosophers), then this Intellect (alone) gave rise to the world. In one sense this may be true for Ibn 'Arabî, but he adds – as we have just seen at the beginning of this section – that Allah has a unique Face specifically turned to every single entity in the world, through Which its existence is preserved [II.434.18]. But in that case the philosophers are contradicting – or at least failing to illustrate the relevance of – their own proposition (Ibn 'Arabî also discussed these views in *al-Durrat Al-Baydâ*: 142–143), because again the world is many and the First Intellect is one; so we still need then to explain how this multiplicity of the world appeared out of the First Intellect.

In Ibn 'Arabî's view, however, every individual entity in the world always has a direct creative relation with Allah, and that is how its existence exists and is maintained. If Allah did not maintain this creative 'special face' between Himself and each entity, it would cease to exist instantly (*Al-Durrat Al-Baydâ'*: 133). In order to solve the problem of unicity–multiplicity relation, Ibn 'Arabî actually asserts that this interface between the One and all the many existent things does not happen all at once. Rather, at any instant, as we have just seen, there is in reality a single relation or interface – a unique divine 'with-ness', as he calls it (following the Qur'an) – between the One and each 'one' of the entities of the world. But what happens at this particular instant with the other entities in the world, since their existence is also preserved only through this unique creative relation between them and their Creator, the unique One? The answer is: they do cease to exist, and then they are (immediately) *re-created* again and again [II.385.4]. We shall discuss this central metaphysical principle of the 'ever-renewed creation' in section 5.6.

Therefore, in order to understand the relation between the unique Oneness of the Real and the multiplicity of the creatures, Ibn 'Arabî adds time to the previous philosophical statement, which can be then reformulated as: 'from the One there can emerge only one *at a time*'. This re-statement is indeed the key to understanding Ibn 'Arabî's unique views of time and the oneness of being and to solving the mystery of the relation between the Real and His creation. In this way the world is created by Allah 'in series' (*Al-Durrat Al-Baydâ'*: 139), and not just one single time, just as the repeated images of a movie are displayed on the television or computer screen.

Ibn 'Arabî, moreover, affirms that this particular mode of creation was chosen by Allah to be like that, although in fact He might have done it in any other way, so it is not an (external) restriction over Him:

> So this is not necessarily implied by the Existence of the Real: i.e. that for example only one can emerge out from Him, and that this is impossible (otherwise). But He willed that and He wished it, and if He had wished that the world should exist all at once, and that nothing were dependent on anything (else), it would not be difficult for Allah [to make it like that, and in this case – if Allah had wished that the world should exist all at once – then we would be living in a different logic, but because Allah created it in this way as it is now (ruled by the laws of causality: see also section 7.7), we observe that from the one nothing might emerge except one at a time, since otherwise this would violate the oneness of the Real according to our current logic].
>
> (*Al-Durrat Al-Baydâ'*: 139)

The meaning of this principle is in fact derived directly from the well-known verse in the Qur'an that we have already discussed in many earlier contexts:

each Day He is upon some (one, single) task (55:29). Ibn 'Arabî quotes this verse most frequently in his discussion on time, and it is the basis of his unique 'quantization' of time. So since Allah is One, He does only one single creative task each 'Day' – of course not this normal observable day that we encounter, in which an almost infinite number of tasks or events are happening.

5.3 The unicity of God and His names

According to Ibn 'Arabî and to Islam, and also some other religions, Allah is both One (*Wâhid*) and 'Unique' (*al-Ahad*, 'the Unit').[9] The first attribute means that He is one God (not many) and the second attribute means that He is not divisible into other entities. Moreover, as is indicated by the more metaphysically problematic second attribute of 'Unicity' (*ahadiyya*), we can not describe Him as a single entity (like other entities) with specific dimensions that are placed somewhere in space or began at a point of time. Allah is simply uniquely different from whatever we may know or imagine. We can not achieve full knowledge or awareness of the Essence (*Dhât*) of Allah because this is beyond our perception. It is, however, possible to describe Him and speak about His Attributes and divine Names, for example as they are mentioned in the Qur'an and the Sunna. We may attain knowledge about the divine Names and descriptions of Allah, but not about His Essence (*Dhât*) Himself. As Chittick has pointed out, for Ibn 'Arabî: 'God is known through the relations, attributions and correlations between Him and the cosmos. But the Essence is unknown, since nothing is related to It' (*SDK*: 62). Whatever the human being may know about Allah is therefore in the end partial and incomplete. No one can ever achieve full knowledge or awareness of Him; the maximum knowledge one might achieve of the Essence of Allah is to know that He is different from anything [IV.301.17]. Ibn 'Arabî expressed this nicely in his prayers: 'it is enough for me that You know my ignorance. You are as I know, and beyond what I know to a degree that I do not know' (Beneito and Hirtenstein 2000: 131–132)[10] This is because we may know Him only through His manifestations in us and in the world, but His manifestations are never exactly repeated.

Human Being, however, is the creature most capable of knowing Allah, the Exalted, because when He created Adam (the Perfect Human Being), He *taught him all the Names* (2:31)[11] and ordered the angels to prostrate before Adam out of respect and acknowledgement (2:34, 7:11, 17:61, 18:50, 20:116, II.46.33). But knowing Allah is an infinite process for us, because Allah Himself is not finite, in the sense that He never manifests in the same form twice [I.266.10], and also because His manifestations reveal some of his attributes and descriptions, but do not fully reveal His ultimate Essence or Identity. Ibn 'Arabî summarizes this by saying, in one of his many elaborations of the famous Divine Saying of the 'Hidden Treasure', that:

> Allah, the Exalted, 'loved to be known' in order to grant the world the privilege of knowing Him, the most Exalted. But He knew that His Identity

> (Essence) can not be (completely) known and nobody can ever know Him as He knows Himself. The best knowledge that can be achieved about Him, His Highness, in the world is that the knowers know that they do not know. And this (human inability to know the Essence) is (also) called knowledge, as the Righteous (Abû Bakr al-Siddîq) said: 'The incapacity to attain realization is a realization.'[12]
>
> [III.429.7]

As Ibn 'Arabî notes elsewhere, the Prophet Muhammad has also clearly expressed this same recognition by saying: 'I can not enumerate the ways of praising Thee: Thou art as Thou has praised Thyself' [*Kanz*: 2131, 3652, I.126.15, I.271.5, etc.], and Allah also said: '*but they do not encompass Him with knowledge*' (20:110).

Of course the Prophet Muhammad also said: 'Allah has 99 Names, one hundred less one: whoever enumerates them is going to enter the Garden' [*Kanz*: 1933, 1934 and 1938]. However, this does not *restrict* the divine Names to 99, as some Muslims misunderstand it. In fact Ibn 'Arabî says that the divine Names of Allah are countless and that everything in the cosmos is a divine Name. But the 99 enumerated Names are the principal Names [IV.288.2] that have been mentioned in Qur'an and Sunna. Moreover, although each Name of the divine Names is different from others, Ibn 'Arabî repeatedly cautions his readers that all that Names are intrinsically implicit in each one of them, which is to say that each Name can be described by all the other Names [I.101.5]. However, despite the multiplicity of these Names, they all refer to the same One Absolute Essence of Allah, while conveying different Attributes of Him due to His manifestations and relations [I.48.23]. Multiplicity is not an intrinsic property of Allah Himself, since Allah has many different Names only when considered with relation to His creations:

> The Names of the Real do not become plural and multiple except in manifestation. But with respect to Him, the property of number does not rule over them, not even its (the number's) root, which is (the number) one. So His Names, in respect to Him, may not be (exclusively or restrictively) described by unity or multiplicity.
>
> [II.122.19, see also *Al-Masâ'il*: 109]

So in fact even the Names 'the One', 'the Unique' and the like are not descriptions of Allah with respect to Himself, but with respect to his creation. If we suppose that there is no creation, there would be no need to describe Him by the One or any other Name. Ibn 'Arabî frequently points out that this is just like the fact that the meaning of the number one is only introduced with regard to its relation to the other numbers.

So Allah may be known only through His divine Names. And because these Names are countless, our knowledge of Allah may never be complete. It is these knowable divine Names that are actively engaged and manifested in the creation.

That is why we see multiplicity and diversity in the cosmos.[13] In fact Ibn 'Arabî maintains that everything in the world is in essence a divine Name, simply because everything is a cause that we need, and Allah says: '*you are in need of Allah*' (35:15) (III.208.7; see also section 7.7). Also He said that '*He is the First and the Last and the Manifest and the Hidden*' (57:3). He manifests in all things, so the things are not other than Him; but also the things are not (identical with) Him, as we shall explain shortly.

However, we should take all these descriptions and names as mere approximations, because they are words spoken in our own language: the names (words) that we know are actually *the names of the Names*, and not the Names themselves [II.56.33]. Although we may know about Allah by knowing His Attributes and Names, those outward verbal Names are words in our language so that we may, for example, look up their meanings in the dictionary, or even use them to name and describe people and things. So although those same familiar words are Names of Allah, their actual meanings are quite distinct when Allah is called by them. For this reason Allah is named as *al-fard* ('the Singular'), because He is distinct (or 'singled-out': *mutafarrid*) from the creation [IV.276.33]. Also, all His Names are described by their 'singular uniqueness' (*al-tafarrud*). As Chittick has pointed out, those words that are revealed to us (through the Qur'an and Sunna) are the outward forms (*sûra*), while Allah's own knowledge of Himself is the reality or inner meaning (*ma'nâ*) (*SDK*: 34). Similarly the Names that are revealed to us in everything in the cosmos are the outward forms, while the inner meaning of those forms is Allah's own knowledge of Himself. Ibn 'Arabî showed that:

> Allah says: '*Call upon Allah or call upon the All-Merciful: whoever you call upon, to Him belong the most beautiful Names*' (17:110). So here He made the most beautiful Names belong to Allah as they (also) belong to the All-Merciful.
>
> But here there is a subtle point: since every Name has a meaning (*ma'nâ*) and a form (*sûra*), 'Allah' is called by the Name's *meaning*, while the 'All-Merciful' is called by the Name's *form*. This is because the (divine) Breath is ascribed to the All-Merciful,[14] and through this (creative) Breath the divine words become manifest within the levels of the Void (*khalâ'*) where the cosmos becomes manifest. So we call Him only by the form of the Names (and only He Himself knows the real meaning of these names).
>
> [II.396.30]

Therefore whatever knowledge we may acquire about Allah actually belongs to His Name 'the All-Merciful', and the difference between the two Names is like the difference between the form and the meaning. According to Ibn 'Arabî, this relation is a direct implication of the verse in Qur'an: '*the All-Merciful mounted on the Throne*' (20:5). In the first chapter of *Al-Tadbîrât Al-Ilâhiyya* (p. 89), which talks about the Universal Spirit (*al-Rûh al-Kullî*) – which is the 'Greatest Element/Single Monad', as we shall see in sections 6.2–6 – Ibn 'Arabî makes a comparison between the Names Allah and the All-Merciful. Under a section called: 'a secret for the special (people of Allah)', Ibn 'Arabî says that the difference between those two Names is like the difference between this Universal Spirit on which Allah mounted (or established His authority) and the Throne on which the All-Merciful mounted (or established His authority). We shall see in section 6.6 that the same comparison can be made between the Single Monad and the Greatest Element. Just as the divine Throne encompasses all the cosmos, everything in the cosmos is related to the Name 'the All-Merciful' [II.467.21].

5.4 The metaphysical triplicity ('trinity') of the cosmos

According to Ibn 'Arabî, as we have already mentioned in section 3.6, the divine Names can be grouped into three categories: the Names of Essence (*asmâ' al-dhât*), Names of Descriptions or Attributes (*asmâ' al-sifât*) and Names of Actions (*asmâ' al-af'âl*) [I.423.20, I.67.28, see also *Inshâ' al-Dawâ'ir*: 22]. Therefore the Universal Intellect or Adamic 'Perfect Human Being', who is created 'according to the Image of the All-Merciful' [*Kanz*: 1146, 1148, 1149], came out with three faces because he is based on these three unique divine dimensions [I.446.19, II.434.16], although he is in essence one indivisible entity.

That is why Ibn 'Arabî asserts that (everything in) the cosmos is built on a kind of metaphysical triplicity [III.126.21], and he later explains this further by saying:

> So the body is (composed of at least) eight points, just as the knowable (aspects) of the Real are the Essence and the seven Attributes: They are not Him, and They are not other than Him. (Likewise) the body is not other than the points, and the points are not other than the body. Now we only said that eight points are the minimum (required to compose) bodies because the name 'line' is for two points or more. And the origin of the plane is from two lines or more, so the plane is from (at least) four points. And the origin of the body is from two planes or more, so the body is from eight points.
>
> Therefore the name (attribute) of 'length' is applicable to the body from the line (included in it); the name 'width' is applicable to it from the plane; and the name 'depth' is applicable to it from the combination of two planes. Thus the body is built on a triplicity (of dimensions: *tathlîth*), just as the formation of proofs (in syllogistic logic) is based on a threefold structure, and just as the Source of existence – that is, the Real – only becomes manifest through the bestowing of existence through three realities: His Entity, His willing intention (*tawajjuh*) and His Speaking (the Command 'Be'). Thus the world became manifest 'according to the Form' of the One Who gives it existence, both in sensation (i.e. the visible world) and in (its spiritual dimensions of) meaning.
>
> [III.276.1]

For this reason – i.e. the threefold structures underlying all generative processes – Ibn 'Arabî emphasizes that we need two elements (subject and object) in order to produce a result (act) [I.278.14], because 'from the one alone nothing may be produced' [III.126.1].

In fact the metaphysical triplicity of subject, object and resulting act is fundamental throughout Ibn 'Arabî's cosmology. Thus he summarizes the process of creation by saying that the Universal Intellect (subject or symbolic 'father') is writing down in the Universal Soul (the object or 'mother') all that Allah wants to create till the Last Day (i.e. the result or 'son') [*Ayyâm Al-Sha'n*: 7–8] (see section 4.3). Indeed Ibn 'Arabî even wrote a separate treatise dedicated to explaining certain threefold structures of expression and meaning in the Qur'an.[15] It is noteworthy in this regard that the metaphysical concept of triplicity is also fundamental in many other ancient religions and philosophies, including the Babylonian, Persian, Egyptian, Roman, Japanese and Indian.[16]

For this reason, Ibn 'Arabî regards the number three as the first 'single' (or odd: *fardî*) number, because he consistently considers that one is not a number (since the Arabic word for 'number', *'adad*, implies multiplicity), while two is the first number. But since the number two is even, nothing may be formed without a third, intermediate principle to link or interface between the one and the two:

> So no contingent entity (*mumkin*) has come to exist through one (alone), but only through a plurality (or 'conjunction' of elements: *jam'*), and the least of plurals is three.[17] So since the (divine) Name 'the Singular' (*al-fard*) is threefold in its effects, He gives to the contingent entity that He brings into existence those three things (i.e. Essence, Will and the creative Command) that He must unavoidably consider, and (only then) He brings (the contingent thing) into existence.
>
> . . . Therefore this (metaphysical principle of) triplicity runs inwardly through the totality of all things, because it exists in the (divine creative) Source.
>
> [III.126.5]

Now the Universal Intellect is the intermediary reality (also called *barzakh* or 'connector') between Allah and the world, and therefore It has two interfaces: It faces Allah from the side of His unity, and It faces the world from the side of His threefold creative dimensions. When It faces Allah (in order to perceive knowledge), It turns away from us (the world), and this is our 'night'; and when It faces the world, this is our 'daytime' or manifest existence. The Perfect Human Being or First Intellect is – according to the famous hadith already discussed above – the 'Image' or 'Form' of the Real, and likewise the world is the subsequent image of this Perfect Human Being. Hence the manifest world, like the Perfect Human Being, is 'according to the Image' of the Real Himself – although without the Perfect Human Being it could not participate in this perfection (see section 3.1). So if we consider the manifest images of the world, we can potentially discover the face of the Perfect Human Being reflected in it, and if we come to know the Perfect Human Being, then we also come to know Allah.

So this is the fundamental ontological triplicity of 'Allah-Human Being-world' or 'Allah-Intellect-world' [I.125–126].

According to Ibn 'Arabî, this fundamental triplicity of Allah-Human Being-world is manifested again within each human being as in our threefold nature as spirit-heart-body. The spirit is the single immaterial and mysterious divine reality that is the principle underlying life and creation, while the body is the place where this creation occurs in many different ways, so it is composed of many different material parts (multiplicity). And the heart is the link between the body and the spirit through which the spirit exerts its effects on the multiplicity of the body. On the other hand, the sensations collected by the body are eventually raised up to the spirit as spiritual (immaterial) meanings and realities.

Another favourite symbolic triplicity for Ibn 'Arabî, deeply rooted in the symbolism of the Qur'an and certain key hadith, is the astronomical triplicity of the Sun–Moon–Earth.[18] The Earth gains its life from the Sun, and when the Sun does not face the Earth from a specific direction, the Moon takes part and reflects the light of the Sun with a degree that is small or large according to its relative place in space. In fact, if one wants to look at the Sun, one is instead obliged to look at the Moon when it is full, because the Sun can not be seen unveiled at all, since it will burn up everything that its direct light falls on. This symbolism is directly connected, for Ibn 'Arabî, with the famous 'hadith of the veils', according to which Allah has 70,000 veils of light and darkness, such that if He removed those veils His light would burn everyone who tried to see Him directly [*Kanz*: 29846, 39210].

Indeed the trinity of Sun–Moon–Earth particularly well illustrates Ibn 'Arabî's view of the creation and its relation to the Creator. Although the creation by Allah is done 'through' the Universal Intellect, Ibn 'Arabî also emphasizes that Allah also has a *direct*, 'individual face' turned toward every single entity in the world. Similarly the Sun does not only give its light indirectly through the Moon, but also much more directly to the Earth, so everything on Earth is connected with the Sun in the course of the day with different degrees and at different times.

5.5 The oneness of being

The 'oneness of being' (*wahdata al-wujûd*) is one of Ibn 'Arabî's most controversial doctrines which many later Muslim scholars attributed to him, usually with very different (and often more polemic than philosophical) meanings and interpretations. Although Ibn 'Arabî himself never mentioned the precise term '*wahdata al-wujûd*' in his writings,[19] it is quite evident that Ibn 'Arabî's books are full of statements that develop notions related to the oneness of being in one way or another, in many places quite explicitly and rigorously. This is especially the case in his most controversial book, *Fusûs al-Hikam*, for which he was widely criticized, but related discussions are also to be found throughout the *Futûhât* and his other shorter works. Indeed the possible misunderstandings of this conception clearly underpin Ibn 'Arabî's distinctive multi-layered, intentionally 'scattered' rhetoric and writing style throughout the *Futûhât* and other works, as he explained quite clearly in the key Introduction to the *Futûhât* itself.

The basic ontological issue for Ibn 'Arabî is very clear and simple: in many places throughout his writings, such as the long chapter 198 of the *Futûhât* [II.390–478] he follows the established Avicennan distinction, familiar to all students of Islamic theology and philosophy by his time, in dividing all conceivable things, in terms of existence, into three basic categories (see also section 2.3). There he says:

> Know that the matter (of the nature of the reality) is the Real (*al-haqq*) and the creation (*al-khalq*): that is the absolute Existence that has always been and always is (existing); and absolute (contingent) possibility (*imkân*) that has always been and always is (possible to exist); and absolute non-existence that has always been and always is (non-existing). Now the absolute Existence does not accept non-existence, (and that applies) eternally and perpetually. The absolute non-existence does not accept existence, (and likewise that applies) eternally and perpetually. But the absolutely possible does accept existence through an (ontologically determining) cause, and it also accepts non-existence through a cause – and (that contingent ontological status also applies) eternally and perpetually.
>
> So the absolute Existence is Allah, nothing other than Him. The absolute non-existence is the impossible-to-exist, nothing other than it. And the absolutely possible (of existence) is the world, nothing other than it: its (ontological) level is between *the* absolute Existence and absolute non-existence. So in so far as some of it faces non-existence, it accepts non-existence; and in so far as some of it faces Existence, it accepts existence. So some of it is darkness, and that is the Nature. And some of it is light, and that is the 'Breath of the All-Merciful' [that is the 'real-through-whom-creation-takes-place' (*al-haqq al-makhlûq bihi*), see sections 6.3 and 6.5] which bestows existence upon this possible (realm of created beings).[20]
>
> [II.426.26]

After that Ibn 'Arabî explains the different types of Creation according to this creative Spirit – or the divine Name 'Light', which he often uses as synonymous with the divine creative and existentiating power – is attached to the 'dark', contingent form of the creatures. Then he goes on to gives the crucial analysis which clearly explains his profound view of the oneness of being in the most explicit and direct way, based on evident verses in the Qur'an. He says:

> So the possible (contingent) existence became manifest between light and darkness, nature and spirit, the unseen and the visible, and the 'veiled' and unveiled. Therefore that which is close to (*waliya*) absolute Existence, from among all that (contingent realm) we have mentioned is light and spirit, and all of what we have mentioned which is close to absolute non-existence is

'shadow' and body – and from the totality (of those different kinds of contingent existent) form (*sûra*: of the whole of creation) comes to be. So when you consider the world from the side of the Breath of the All-Merciful, you say: 'It is nothing but Allah'. But when you consider the world with regard to its being equally balanced and well-proportioned (i.e. between existence and non-existence), then you say these are creations (*makhlûqât*). So [in the famous Qur'anic expression of this fundamental ontological reality, addressed to the Prophet]: '*you (Muhammad) did not throw*', inasmuch as you are a creation [so that it is God who was really acting], '*when you did throw*', inasmuch as you are real (*haqq^{an}*), '*but Allah threw*' (8:17), because He is *the* Real (*al-Haqq*).

For it is through the (divine creative) Breath that the whole world is 'breathing' (animated with life), and the Breath made it appear. So (this creative divine Breath) is inner dimension (*bâtin*) for the Real, and the manifest dimension (*zâhir*) for creation: thus the inner dimension of the Real is the manifest dimension for creation, and the inner dimension of creation is the manifest aspect of the Real – and through their combination the generated existence (*al-kawn*) is actualized, since without that combination it would (only) be said to be Real and creation. Thus the Real is for the absolute Existence, and creation is for the absolutely possible. So what becomes non-existent of the world and its form that disappears is through what is close to the side of non-existence; and what remains of it and does not allow for non-existence is through what is close to the side of Existence. Hence these two things (Existence and non-existence) are continually ruling over the world, so the creation is always new with every Breath, both in this world and in the hereafter.

Therefore the Breath of the All-Merciful is continually directed (toward the Act of creation), and Nature is continually taking on existence as the forms for this Breath, so that the divine Command does not become inactive, because inactivity is not appropriate (for It). So constantly forms are newly appearing and becoming manifest, according to their states of readiness to accept the (divine creative) Breath. And this is the clearest possible (description) of the (divine) origination (*ibdâ'*) of the world. *And Allah says the truth and He shows the way* (33:4).

[II.427.17]

To summarize, therefore, this expression implies that the world can be conceived symbolically as a mixture of light and 'darkness'. For Ibn 'Arabî, this darkness is quite literally nothing: it is simply the absence of light.[21] Light, on the other hand, is ultimately the Real (via the divine Name 'The Light', *al-nûr*), and the Real is One. So all existence is in essence one. Multiplicity appears through creation as a result of mixing the oneness of light with the darkness of non-existence. In other words, we can say – since darkness is nothing – that the creation is the constantly repeating relative appearance (manifestation) of the Real. The Real manifests most perfectly in the Perfect Human Being, and relatively in other creatures, and these manifestations happen through the Universal Intellect. So in real existence there is only the Real Who is Allah and this Universal Intellect who is the Messenger of Allah. So there is in fact no ontologically self-subsistent 'evil', since the creation is all good – an aspect of this cosmological conception which, taken out of context, could easily give rise to obvious religious and ethical objections. Hence what we perceive as evil is in reality the absence of good, just as darkness is the absence of light.[22]

Thus this is the basic principle, but in order to understand it we need to explain how the mixing between light and darkness is done, which is again to say: 'how is the world created?' – which raises the question of time yet again. Ibn 'Arabî's understanding of this process of creation or cosmogony will be developed further below and in sections 9.7–8.

Given the possible confusions and misunderstandings surrounding this understanding of creation, it is clear why Ibn 'Arabî never declared these ideas in overly simplistic terms in his books, but rather scattered them throughout his writings – as he explains quite explicitly at the very beginning of his *Futûhât* [I.38.25] – so that the common people would not misunderstand them (as indeed happened in later times) and so that only those properly 'prepared' would be able to discover their profound intended meanings.

Moreover, we have to admit that Ibn 'Arabî takes it a courageous step further: although the Universal Intellect, and hence the entire manifest world, is created by Allah, Ibn 'Arabî emphasizes that it also is not 'other than Allah' (*SDK*: 113), because ultimately only Allah has real absolute and necessary (independent) existence [I.194.8], and the world exists by and through Him not by itself. As Chittick has summarized this perspective (*SDK*: 81), if Ibn 'Arabî was asked the question: 'Are the things the same as God?', his answer would be: 'Yes *and* No'. That is to say, they are, in his own words, 'He/not He' (*Huwa lâ huwa*);[23] or equally, one could say: 'they are not Him, and they are not other than Him' (*Lâ hiya huwa wa lâ hiya ghayruh*).[24] For if we say 'Yes' (alone), then this would require us confining Allah, the most Exalted, in objects, which is an obvious misconception. And if we say 'No' (alone), then this would require the assertion of other separate and self-subsistent) existents, and this – for Ibn 'Arabî – is also wrong. So the ultimate truth requires combining both ontological views and saying that the things are in essence 'not other than Allah' – although in the forms that we see, they also are not (identical with) Allah. These forms do not have real independent existence, since otherwise Allah would not be 'the One (alone)' (*al-wâhid*) – but He is the One (alone), and the created things exist by and through Him, not by themselves. For Ibn 'Arabî, this is in fact 'the secret of sincerity' (*sirr al-ikhlâs*), which is also 'the secret of destiny' (*sirr al-qadar*) that makes clear the fundamental distinction between the Creator and the creation, the Eternal and the created – and this secret, he explains, has been hidden from most people [III.182.11].

Chittick goes on to show that, for Ibn 'Arabî, everything in the world is a divine name of Allah (*SDK*: 94); or rather, the things are the manifestations of the Names. From the general Qur'anic verse *O people, you are truly in need of*

Allah (35:15), Ibn 'Arabî easily concludes that everything (that is a cause, and everything is a cause) is a divine Name, because we are in need of these causes, so they may not be other than Him [I.288.16]. Again we can not say that these secondary and intermediate causes (*asbâb*) *are* Him – though in their most ultimate essence and Source they are all one, and this one is not other than Him. But of course someone simply hearing such an expression will initially think that they mean that the particular shape or body of the created objects that are the causes: that is why we can not simply stop at saying (as many later Sufi poets sometimes did) that 'all things are Allah'. Ibn 'Arabî explained these key ontological distinctions most clearly and extensively in his *Fusûs al-Hikam*, especially in chapter 3 'on the wisdom of transcendence (*al-hikma al-subûhiyya*) in the word of Noah';[25] but he also took up this same subject at many places throughout the *Futûhât* [I.90.17, and the extensive references cited in the previous paragraph].

In a similar manner, Ibn 'Arabî often describes the world of all creation as a kind of 'mirror' on which Allah's Image is reflected [IV.430.1]. If someone looks at the world from the real side of actual existence, then he will see the Image of the Real, Allah the most Glorious; but if he considers the world only from the side of its non-existence (if we suppose this is possible), then he will see an image of the unreal:

> Therefore the reason why this (ontological) 'isthmus' (*al-barzakh*) – which is the possible (realm of contingent existence) between (pure) non-existence and Existence – is the occasion (*sabab*) for its being attributed both permanence and non-existence, is because it corresponds to both those things by its essence. That is because the absolute non-existence stood up like a mirror for the Absolute Existence, so the (divine) Existence saw His Image in it, so that this Image is the essential reality (*'ayn*) of the possible. That is why this 'possible' (as the Perfect Human Being or First Intellect) had a permanent individual-essence (*'ayn thâbita*) and state of (definable) 'thingness' already in the state of its non-existence. And that is why it emerged (in its contingent, created existence) according to the Image of the Absolute Existence. This is also that is why it was also describable as non-finite, so it is referred to as infinite.
>
> But the Absolute Existence is also like a mirror for the absolute non-existence. So the absolute non-existence saw itself in the mirror of the Real, but the image that it saw was itself the essential reality (*'ayn*) of non-existence by which this possible (existence) is described. Therefore it is also described as infinite, just as the absolute non-existence is infinite – hence the possible is described as (inherently) non-existing (*ma'dûm*). So it is like the image that appears between the mirror and the person looking in the mirror: that image is not that very person himself, but it is not other than him. Likewise the possible, with respect to as its (very limited kind of) permanence, is not the very essence of the Real Himself; yet it is not other than Him. Similarly, with respect to its (only relative) non-existence, it is not the

> same thing as of the (absolutely) impossible, yet it is not entirely other than it. So it is as though it is something relative (depending on how it is viewed).
>
> [III.47.32]

He also says:

> If the unreal (non-existence) had a tongue (to speak), it would tell you: 'you are according to my image', because it sees in you nothing but its own shadow, just as the Existence has speech and has said: 'you are (created) according to My Image' [*Kanz*: 1141–1150, and 15129], because He saw in you His own Image.
>
> [IV.154.23]

So the world may not have a constant real (self-subsistent) existence, because only Allah may be described by that; and at the same time the world is not in constant non-existence, or it would not be there at all. Instead it is perpetually fluctuating, at every instant of creation, between existence and non-existence. And in fact, Ibn 'Arabî points out [II.303–304], this is the real meaning of the Qur'anic symbols of the 'daytime' and the 'night': when the Universal Intellect (Allah's Messenger) faces the Real, this would be a kind of 'night' for us, but when the Intellect/Messenger faces us – in each divine Act of creation – this is our manifest day (*shahâda*: what is 'seen' or manifest – see also the related discussions in sections 2.14 and 6.8 [Spreading the Shadows]). What makes sense of this distinction, of course, is Ibn 'Arabî's assertion of the central cosmogonic principle of perpetual re-creation, which we shall discuss next. From that perspective, the world is actually continually created 'in series', bit by bit, one entity at a time; so no two entities may gain real existence at the same time, because they gain their existence only through their constant re-creation by the Real who is One.

For Ibn 'Arabî, this rule or principle of serial creation and re-creation is in fact a direct implication of the verse '*each day He is upon some task*' (55:29); at the same time, it is for him the actual underlying meaning of the familiar philosophical maxim 'from the one there might proceed only', which we discussed in section 5.2. We have seen, however, that for Ibn 'Arabî this creative procedure is not therefore imposed on the Real; rather, it remains His choice which He makes when He creates the world.

We also must remember that this divine oneness manifest in the Act of creation that we can know and observe – which grounds our own experience of reality and the cosmos – is clearly rooted in the Single Monad. Yet this Single Monad (the First Intellect, 'Muhammadan Reality', etc.) is itself but one of an unknown number of divinely 'originated' (*ibdâ'*) 'Roaming Spirits' (see section 1.4), so there is indeed some kind of (for us) further unknowable multiplicity on this higher ontological level. However, it would appear that again we can apply the same judgement to those Roaming Spirits, according to our own human logic, that they may not be 'other than' Allah, nor the same as Him (since He originated them): so they also are in this ontological status of 'He/not He', which again indicates an ultimate Oneness on the highest divine level.

We have actually already summarized the basic themes of this section right at the end of section 2.1, when we recalled Ibn 'Arabî's conclusion that 'time is defined by motion, and motion is defined by the different positions of the formable monads, and those monads are different states (times or instances) or forms of the Single Monad, that alone has a real existence'. Therefore, 'there is no god but Allah' and this highest created entity (i.e. 'the Single Monad') is the (all-encompassing) Messenger of Allah. Allah sent this Messenger '*to take us* (and the whole created world) *from the darkness*' of non-existence '*into the light*' of existence (Qur'an 57:9). But although this Single Monad is not Allah, he is not other than Allah: he is some sort of fundamental manifestation of Allah, and it is through this primordial manifestation that Allah can accurately and meaningfully be described, in the famous Qur'anic expression, as '*the First and the Last, the Manifest and the Hidden*' (57:3).

5.6 The principle of ever-renewed creation

Ibn 'Arabî's conception of time is profoundly rooted in one of the most famous and distinctive – and uniquely and problematically experiential – features of his world-view, the principle of the 'ever-renewed creation' of all the manifest worlds at every instant. Thus he affirms, in a more abstract statement of this perception:

> There is no doubt that the 'accidents' [i.e. the particular forms taken by creation in all the different levels of existence at each moment] become non-existent in the second instant-of-time after the instant of their coming into existence. So the Real is continuously watching over the world of bodies and the higher and lower (spiritual and imaginal) substances, such that whenever a (particular) form through which they exist becomes nonexistent, He creates at that same instant another form like it or opposed to it, which (new creation) preserves it from non-existence at every instant. So He is continuously creating, and the world is continuously in need of Him.
>
> [II.208.27]

He also makes it clear that this continuously renewed 'return to non-existence' is an intrinsic condition of all the created forms, and not due to any external force [II.385.4]. Typically Ibn 'Arabî relates this fundamental insight to the Qur'anic verse: '*but they are unaware of the new creation (khalq jadîd)*' (50:15), which he frequently quotes[26] – along with the famous verse concerning the 'Day of the divine Task' (55:29) that he cites in relation to his intimately related concept of the quantization of time.

Therefore the existence of things in the world is not continuous, as we imagine and observe, because Allah is continuously and perpetually creating every thing whatsoever – at every level and domain of existence – at every instant, or in every single 'Day of event' [II.454.21, II.384.30]. This means that, just as time (for Ibn 'Arabî) may exist only at one atomic instant at a time, so also space (and whatever it may contain) also exists only one instant at a time. In fact there is no difference between space and time: they are both containers for events (see section 2.1 and 3.6).

We have already seen in previous chapters that, according to this principle of ever-renewed creation, Ibn 'Arabî explains motion (section 2.6) in a new unique and unprecedented manner, and he also explains the 'intertwining' of days (section 4.4) as well as some other related philosophical and theological concepts. Indeed this hypothetical (though Ibn 'Arabî affirmed it through unveiling) principle forms the basis of Ibn 'Arabî's overall view of the world, and we shall use it as one of three key hypotheses in the following chapter in which we explain his Single Monad model of the cosmos.

Of course this re-creation must be happening at extraordinarily high 'rates of refreshment', but Ibn 'Arabî has no difficulty in finding scriptural allusions and theological and other arguments supporting this distinctive conception, in addition to the direct experiential evidence of the spiritual 'knowers' (*'urafâ'*). He says:

> The form becomes non-existent in the next instant-of-time after the time of its coming into existence, so the Real is always Creator, and the monad (substance, *jawhar*) is always in need (of the Creator for its existence). For if the form would remain (for two instants of time or longer), those two principles would not hold. But this is impossible [at least theologically speaking, since otherwise the creatures would be independent of Allah, whereas only Allah maybe described as (completely) Self-sufficient (*al-Ghanî*), while everything created has the essential intrinsic quality of ontological 'poverty' (*faqr*) or need of the Creator for its very existence], so it is impossible for the (created) form to remain for two instants of time (or longer).
>
> (*Al-Tanazzulât Al-Layliyya*: 55)

So from the theological point of view, neither the forms nor the essences of the created world may remain (constant) for more than one moment because if they do they would be independent of the Creator – whereas both the essence and the 'accidents' or form of the creatures are always in need of their Creator. The essence needs ever-renewed forms because it exists only when it wears a form; and the form does not stay the same, because, if it did so, the Real would not be Perpetually-Creating (*Khallâq*), and the individual form would be at least partly independent of the Real.

In addition, Ibn 'Arabî argues in similar theological terms that there are never any two truly identical forms, since otherwise Allah will not be described as 'the Infinitely Vast' (*al-Wâsi'*). But because of this unique divine Vastness (*al-ittisâ' al-ilâhî*) [I.266.8], the monad will never wear two identical forms: i.e. it never wears exactly the same form for more than one instant; nothing is ever truly repeated [I.721.22]. The new forms, he admits, are often 'similar' to the previous ones but they are not the '*same*' [II.372.21, III.127.24]. Ibn 'Arabî summarized this argument as follows:

> The world at every instant of time (*zamân fard*) is re-formed (*takawwun*) and disintegrated. So the individual entity of the substance of the world (*'ayn jawhar al-'âlam*) has no persistence (in existence) except through its receiving of this formation (*takwîn*) within it. Therefore the world is in a state of needfulness perpetually: either the forms are in need (of a creator) to bring them forth from non-existence into existence; or else the substance [*jawhar*: i.e. the substrate for the created 'forms' or 'accidents'] is in need to preserve its existence, because unavoidably a condition for its existence is the existence of the formation of those (newly re-created forms) for which it is a substrate.
>
> [II.454.19]

Elsewhere Ibn 'Arabî gives a very short statement of this general ontological argument which is rather difficult to follow. There he says:

> The doer (*al-fâ'il*) may not do nothing (although we use this expression widely in our daily language, but it is logical nonsense to say '*do* nothing'), and the thing may not become non-existent (simply) by its opposite, because (the thing and its opposite) may not meet, and (also) because the opposite does not exist (at that same time) . . . So that is why we said that (the form) becomes non-existent by itself (not through an external force or action) and is impossible to remain (in existence for two moments or longer).
>
> (*Al-Tanazzulât Al-Layliyya*: 55, see also I.39.8)

Ibn 'Arabî also made the same basic arguments near the beginning of the *Futûhât*, through the tongue of a mysterious 'western Imâm' who summarizes in an extremely condensed form – usually using rhymed prose and the language and terminology of kalam theology – a series of challenging ontological premises subscribed to by those who have moved beyond the simple creed of the common believers, which forms the second of the four levels of faith that Ibn 'Arabî explained at the beginning of the *Futûhât*, and the mysterious speaker's remarks in this section are roughly based on al-Ghazâlî's intentionally popular kalam treatise *Al-Iqtisâd fî Al-I'tiqâd*:

> (4) Then he said: 'Whatever individual entity (is said to) appear, but which does not give rise to any (distinctive) quality, then its existence is obviously impossible, since it does not give rise to any knowledge (as would be the case with anything that actually exists).'
>
> (5) Then he said: 'And it is impossible for it to fill different places, because its travelling (from one place to another) would be in the second instant of time of the time of its existence in itself, but it does not continue

> to reside (over two instants). And if it were possible for it to move by itself, then it would be self-subsistent and would have no need of place. Nor does (the advent of) its opposite make it cease to exist, because it (the opposite) does not exist (at the same time in the same place). Nor does (another) doer ('make it into nothing'), because although people do use the expression 'doing nothing', no intelligent person maintains that (is possible).
>
> [I.39.7]

These statements are based on two issues: (1) the forms in the worlds of manifestation may not remain longer than a single instant of time, since otherwise they would be independent and self-subsistent; and (2) the created forms intrinsically return to non-existence after every instant. Otherwise they would either be brought into non-existence by the existence of their opposite (which can not exist at the same time and place, and therefore itself has to be newly created), or by another doer who 'makes them non-existent', which is also not logical, because it is nonsense to take literally the expression 'do nothing', where the result of an action is pure non-existence.

Something verbally resembling this notion of the 're-creation' or the perpetual 'recurrence of creation' had earlier been employed by the kalam scholars (MacEy 1994: 47, MacDonald 1927: 326–344) and particularly the Ash'arites who, like Ibn 'Arabî, maintained that the world is composed of substances and accidents (Corbin 1969: 203, Wolfson 1976: 466–517), or monads (substances: *jawhar*) and their forms or 'accidents' (*'arad*). Ibn 'Arabî acknowledged their contribution – and certainly borrowed much of their theological language, giving it his own distinctive meanings – but he also took it a very important step further by saying that even those existing monads are only copies or reflections of the Single Monad or Substance that alone has a real existence [III.404.25, *Al-Masâ'il*: 32]. Therefore, all the forms and monads in all the worlds are continuously and perpetually created and re-created by this Single Monad. Understanding the world therefore requires us to explain and understand how this Single Monad creates the monads and the forms, or in other words how to link the unique oneness of the Creator and the observable multiplicity of the world.

We shall also see in section 7.6 that with the re-creation principle one can easily resolve the standard EPR criticism of the (apparent) inconsistency between Quantum Theory and the theory of Relativity, which arises from our inadequate understanding of the nature of time.

6 The Single Monad model of the cosmos

The cosmos is only imagined, though it is – in reality – real.
And the only one who understands this fact has surely accomplished all the secrets of the path.

(*Fusûs*: 157)

The knowings of the Real are not hidden from anyone,
except for someone who does not know the One/one.[1]
[IV.181.31]

For Ibn 'Arabî, a true understanding of time is the key to realizing and understanding the origin and structure of the world, by providing the link between its ultimate unity and apparent multiplicity. In this chapter we shall outline an integrated cosmological model based on Ibn 'Arabî's view of the oneness of being explained in Chapter 5, together with his understanding of the flow of time as explained in Chapter 4 and his unique view of creation in the Week as explained in Chapter 3. We shall call this model the 'Single Monad model'. Under this model, Ibn 'Arabî views the entire created world, both spiritual and manifest, as imaginal forms perpetually re-created by the Single Monad (*al-jawhar al-fard*) (*Al-Mu'jam Al-Sûfi*: 297), which alone has real existence. This Monad continuously and perpetually appears in different forms creating the phenomena of the visible and invisible worlds.

6.1 The three hypotheses

The Single Monad model can be summarized in the following three hypotheses – although, to be sure, Ibn 'Arabî himself presents these concepts primarily not as the subjects of philosophical or theological arguments, but rather as the symbolic expression of the actual metaphysical realities directly perceived by himself and many other realized 'Knowers' (*'urafâ'*, *muhaqqiqûn*, *ahl Allâh* and so on) through the processes of inspired 'unveiling'

6.1.1 The Single Monad

There is only one *Single Monad* that can be said to have a real existence at any given time. This Monad creates other monads by manifesting different forms ('imaging Itself') to make a comprehensive 'still picture' of the entire cosmos. This still picture is created in one full Week of the original creative Days (of events), but this creative process is equivalent only to one single moment (the 'atom of time') for an observer inside the cosmos. Thus the observed cosmos is the eternally renewed succession of these still pictures made up by the Single Monad. The Monad – discussed in detail in the following sections – is an indivisible reality, but it is still itself a compound made up by the 'Greatest Element' that is the only real ultimate substance in existence.

6.1.2 The re-creation principle

The forms of manifestation cease to exist intrinsically right after the instant of their creation, and then they are re-created again by the Single Monad in every original creative 'Week' (i.e. at every moment). As discussed in sections 3.6 and 5.6, this perpetual re-creation happens in the 'six Days' of creation from Sunday to Friday, which accounts for the three dimensions of space. But we do not witness this creation process as such; instead we witness only the created world on the 'last Day' of Saturday. So the seven Days of the divine Week are in all one point of space-time (Days 6 to 1) which then – by repeating manifestations – manifests the space-time container (i.e. 'the Age', as discussed in section 2.19) which encompasses the world both spatially and temporally.

6.1.3 The actual flow of time

Since the world takes seven Days to be created by the Single Monad, which manifests the forms of the monads one by one in specific order, any observer would have to wait – somehow out of existence – six Days (from Sunday to Friday) in order to witness the next moment of creation (i.e. the next 'still picture') on the following Saturday. But of course we do not perceive all this, but rather experience it as a single infinitesimal moment. In each Day of these Days of creation, a corresponding dimension of the world is created. Therefore, the real flow of the actual created time does not go linearly, but rather is intertwined with the observable, normal earthly days in the special – and admittedly rather mystifying – manner that we have summarized in section 4.4, along with Ibn 'Arabî's account of the way in which the observable earthly daytimes and nights are 'taken out' of each other and separated by three daytimes and three nights in order to form the three-dimensional space that we experience.

6.2 The Single Monad

Borrowing his language from the atomist physical theories of earlier kalam theology, Ibn 'Arabî sometimes refers to the created world as being made up of *monads* and *forms*, or in his technical language, of 'substances' (*jawâhir*, s. *jawhar*) and various changing 'accidents' (*a'râd*, s. *'arad*) that inhere in and qualify those substances. In the process of manifestation, the substances appear to remain relatively constant, while the accidents do not stay for more than one moment. In this terminology, the 'monad' or substance (*al-jawhar*) is – as will be discussed shortly – a physical or metaphysical entity that exists by itself, whereas the 'form' or accident (*al-'arad*) exists only through or by some particular monad. The monad, however, may appear in existence only by 'wearing' some form or another [II.179.26], so we do not see the monads but rather only the forms. Also Ibn 'Arabî asserts that the monad exists by itself and its existence is constant and invariable, while the form exists only in the monad and its existence is temporal; it exists only at the time and then it vanishes instantly and intrinsically, and the same never comes back to existence again [II.677.30, III.452.24].

Generally *jawhar* signifies everything that exists in reality. Literally it originally meant 'jewel', but, in this technical sense borrowed from the physical theory of kalam theology, it means 'substance'. In a very different philosophical context, the same Arabic word was used to translate the first of the ten Aristotelian categories already discussed in section 2.1. In English, the word 'monad' – which we will regularly use here for *jawhar* – is derived from the Greek *monados*, and it means 'ultimate, indivisible unit'. It was used very early by the Greek philosophers of the doctrines of Pythagoras, and it was also used later, in a very different way, by the Neoplatonists to signify the One: thus God is described as the 'Monad of monads'.[2]

Like the Neoplatonists, Ibn 'Arabî sometimes uses the term *al-jawhar* (monad) in this higher theological sense to refer to 'the one', 'the essence', 'the real' (not 'the Real' as a divine Name of God, but 'the real-through-whom-creation-takes-place', as discussed further below) and the origin of everything in the world. However, in such cases he does not seem to refer directly to the highest, transcendent dimension of 'God', but rather to the 'Universal/First Intellect' or the 'Pen' [II.675.6], who is also the 'Perfect Human Being'.

On the other hand, although in this theological or cosmological sense the term *al-jawhar* ordinarily refers to the one real essence of the world (of all creation), Ibn 'Arabî also sometimes uses the same term in the plural form (*jawâhir*) to refer to the essences or souls/spirits (*al-nufûs al-nâtiqa*) – or more precisely, to the 'partial intellects' (*al-'uqûl al-juz'îyya*, in contrast with the Universal Intellect, *al-'aql al-kullî*, that is their origin) – of human beings who are the perceivers of the world. Even more generally, he sometimes uses it to refer to any entity (even inanimate ones) in the creation, whether angels, jinn, humans, animals, plants or metals. In this latter more generic sense he considers that everything in creation has a substance which is its monad (*jawhar*) and a particular form (*'arad*) which is its appearance.

These very different dimensions and usages of the term 'monad'/*jawhar*, however, are also intrinsically linked in Ibn 'Arabî's cosmology, since he argues that all the monads of the world are created by, and are therefore the 'images' (reflections, shadows etc.) of the one Single Monad. Thus in that larger perspective of creation, they are nothing but different images of this one Single Monad that in reality may alone be described as having real existence [III.452.24].

6.3 The different names of the Single Monad

As we mentioned above, this Single Monad (*al-jawhar al-fard*) is the Universal Intellect itself and also the Pen. It also has different names or descriptions as Ibn 'Arabî summarized in his book *Al-Durrat Al-Baydâ* (*The White Pearl*) in which he discussed many names and descriptions of the First Intellect and the title of the book itself is one interesting variant. Also Ibn 'Arabî spent much of the first chapter in his book *Al-Tadbîrât Al-Ilâhiyya* on explaining the different names and properties of this Universal Intellect that is the true Caliph (*Khalîfa*). There is, however, some confusion between the Greatest Element that we shall talk about shortly and the Single Monad; sometimes it is not very clear for some of these many variant names whether they are really for the Single Monad or the Greatest Element.

One of these names is 'the real through whom creation takes place' and William Chittick devoted a full section in *SPK* to talk about it (*SPK*: 132). This real-through-whom-creation-takes-place is the most perfect image of the Real Allah, the Creator of the world. That is why he is also called 'the Perfect Human Being'. We do not want to repeat here what Chittick said in *SPK*, but we just want to stress that we should not mix the Real and the real-through-whom-creation-takes-place because they may be confused in Ibn 'Arabî's writings. In this book 'the Real', with capital, is used for the divine Name of Allah (*al-haqq*), while 'the real', with small letter, is used for the real-through-whom-creation-takes-place. But this name actually describes the Greatest Element rather than the Single Monad, because the latter is compound while the Greatest Element is the most elementary 'block' in the world as we shall see shortly. Everything in the Creation at the end is rooted in the real, just as the leaves (and the fruits etc.) of a tree are rooted in the stalk. The leaves were also 'determined' in the seed that gave this tree even before it was planted. So the Single Monad is like the seed for *the tree of the cosmos*,[3] while the real-through-whom-creation-takes-place (i.e. the Greatest Element) is what makes up the seed down to the cells, atoms and subatomic particles inside it.

In chapter 364 of the *Futûhât* Ibn 'Arabî talks about how the entities heard the divine Command in their state of pre-existence and he affirms again what we have mentioned in section 2.3 that for Allah nothing was introduced after His creating the world; only the world moved from determination into existence and from immutable hearing (*sam' thubûtî*) into actual hearing (*sam' wujûdi*). But most importantly, he shows in the following passage that there is in fact only one single entity that has a necessary immutable essence and that is the Perfect Human Being:

> For Allah everything of His servants is still with Him in effect (*bil-fi'l*), nothing is with Him in reality (*bil-quwwa*). So the divine designation came

to him (the servant) with regard to the acts and the states that Allah has with him so that he shall remember with his intelligence what he has already witnessed of his Lord when he was still non-existing – since he had the immutability which made him accept the divine disposition in him, and this state of immutability made him obey the Real's Command regarding the existence because the Command may come only on that who is described by hearing.

So the divine Saying (Command) is still (as It is) and the immutable hearing is still (as it was), but what occurred is only the actual hearing that is like a branch of the immutable hearing. So the state changed on the essence of hearing but the (essence of) hearing did not change, because the essences do not change from one state to another, but the states give them rules so they wear them. But that who has no knowledge imagines that the essence changed. But (the fact is that) the states seek the divine Names and not that the essences are described by seeking. And the essences gain names and descriptions according to the rules of (those) states that change over them. And without the states, the essences would not be distinguished because (in reality) there is only one essence that was distinguished with its entity from the Necessarily Existent (*wâjib al-wujûd*) just as it was described like Him with necessary immutability. So He, the Exalted, has the Necessary Existence and Immutability, and to this essence (only) the necessary immutability. So the states for this essence are like the divine Names for the Real. So just as the Names of the Essence do not multiply the Named nor make Him many, so the states for this essence do not make him multiple or many, despite the rationality of the multiplicity and manyness of the Names and the states. And by this (similarity) it was true for this essence to be described as being on the (divine) Image.

[III.313.35–314.10]

Therefore, in reality, there is only this entity that is the essence of the Perfect Human Being, and the world is the different states of this single entity. As we mentioned at the beginning of this chapter (and also in section 2.1 when we introduced Ibn 'Arabî's view of time, and also at the end of section 5.5 when we discussed the oneness of being), this single entity is the Single Monad, and the essences (i.e. the monads) of the world are different reflections of this Single Monad.

Another important name of the Single Monad is the Universal Spirit (*al-rûh al-kullî*), and Ibn 'Arabî shows that he deserves this name because he goes (v. *râha, yarûh*) in the states of the world:

And this name is to him from two aspects: the first is for his being a spirit, i.e. in ease, happiness and rest (*râha* p.p. of *yastarîh*; to rest or to relieve) owing to his knowledge of his Lord and his witnessing Him. And the second is that he went (*râha* past tense of *yarûh*; to go) through the capacious orbs of the knowledge of his Creator, by a special force. And he went through the

> states of the cosmos to give out to them what Allah entrusted him. And he went through his knowing himself by his need to his Lord and his Creator.
>
> So he has three goings (*rawahât*),[4] so he may be called 'Universal (*kullî*)' Spirit because there is no fourth state other than those to go through. So it is like the imperative (tense) of '*râha*' (رَاحَ: went), '*yarûhu*' (يَرُوحُ: the past participle, to go) and the imperative is '*ruh*' (رُحْ: go!); and when it was transformed from the imperative to the noun, the '*wâw*' was returned to it as also the '*alif*' and '*lâm*' (the definitive article) was added, because the omitting of '*wâw*' from it was due to the meeting of the two consonants.[5] So it is like: he was sought from one direction then it is said that he has gone (*râha*), as we said.
>
> (*Al-Durrat Al-Baydâ'*: 135)

Yet another interesting name of this Single Monad is 'Everything (*kulla shay'*)'![6] This name is interesting because Ibn 'Arabî says in *Al-Masâ'il* that 'in everything there is everything (*kulla shay'in fîhi kullu shay'*), even if you do not recognize that' [*Al-Masâ'il*: 58]. This is on the one hand another expression of his Single Monad theory because it renders into: 'the Single Monad is in everything'. But also it might mean that the internal structure of the Single Monad is as complicated as the world itself because it means: 'in everything (even the Single Monad) there is everything (even the world)!' This last statement is plausible since both the Single Monad (i.e. the Perfect Human Being) and the world are on the divine Image as we have seen before (section 3.2). This reminds us in mathematics with fractals such as the Mandelbrot set, Julia set and Sierpinski triangle, where the structure keeps repeating itself on any larger or smaller scale (Mandelbrot and Frame 2002: 37, Stewart *et al.* 2004: 60). This deserves a separate study, but we just want to mention here that this might answer the question we put forward in section 2.16 about the structure of the moment and whether it is divided into sub-moments. We said there that the moment could be indeed identical to the day where the Sun rises, moves gradually in the sky and then sets to rise again on the next day; as the Single Monad might be identical with the world, the moment might be identical with the day. It just depends on the scale we are using; if we were inside the Single Monad we might see creations such as the Sun, planets and the stars, but because we are outside we see it as a point. Also if we suppose we go outside the world, we shall see it as a point; that is – as the Single Monad – indivisible but compound. This also has its example in modern cosmology as the black hole which occupies a single point in our space but itself is considered a complete world.[7]

6.4 The structure of the monad

At the beginning of the first chapter of *Al-Tadbîrât Al-Ilâhiyya*, Ibn 'Arabî says: 'The first existent originated by Allah is a simple spiritual Single Monad, embodied according to some doctrines and not-embodied according to others' (*Tadbîrât*: 87).

As this remark indicates, Ibn 'Arabî was well aware that there has long been a debate amongst philosophers whether the monad is a physical or metaphysical entity, or whether it is embodied or not [see also I.47.22]. Although he mostly prefers the second choice (*Al-Durrat Al-Baydâ'*: 134), Ibn 'Arabî sometimes does not rule out either case, perhaps because the argument should be meaningless – i.e. the reality must necessarily encompass all manifestations of creation, both spiritual and manifest – if we recall that there is in reality only one Single Monad. Many times, though, he affirms that the Single Monad is embodied and indivisible, especially when the manifest world is concerned [II.438.2]. On the other hand, the essences of the spirits and souls are not likely to be embodied [II.309.25], though both (the manifest and spiritual) are only reflections of the Single Monad that itself can be described neither as (solely) physical nor as metaphysical, because it is necessarily the whole of creation. In the very long chapter 198 of the *Futûhât*, in which Ibn 'Arabî talks in detail about the various aspects of divine creation, he summarizes the various divisions or types of physical and metaphysical entities. He also states the difference between the essences (monads) and their accidents (forms). This is shown in the following long passage in which Ibn 'Arabî also shows the basis of the Single Monad model, while pointing out the difference between the approaches of the Sufis and of the philosophers that we discussed at the beginning of Chapter 5. There he says:

> Now you must know that every knowable thing that may be classified must unavoidably enter into (the category of) what exists in thought (*wujûd dhihnî*). But this thing that exists in thought may belong either to what can receive real existence (*wujûd 'aynî*); or to what may not receive real existence, like the things that are impossible (*al-muhâl*). And that which can receive real existence either subsists by itself, which is called 'not-in-a-substrate' (*lâ-fî-mawdû'*) or else it does not (subsist by itself). And that which subsists by itself is either embodied (or more strictly speaking, 'localized in a place', *mutahayyiz*), or not embodied.
>
> As for that (self-subsistent reality) which is classified as not-in-a-substrate and not embodied, it must necessarily be either what necessarily exists by its own essence (*wâjib al-wujûd li-dhâtihi*), and He is Allah the Exalted; or what necessarily exists through (the determination by) something other than itself, and that is what is contingent (*al-mumkin*: i.e. the whole created world). And this (category of what is) contingent is either embodied (in-a-place), or not embodied. As for the division among the contingent things of what is self-subsistent, that is either not embodied – like the rational souls (*al-nufûs al-nâtiqa al-mudabbira*) that govern the substance of the (spiritual) world of Light, the natural world, and the elemental world, or else (the self-subsistent contingent things) that are embodied are either compound with parts, or without parts. So if it has no parts, it is the (simplest) 'Single Monad';[8] and, if it has parts, it is a (natural, elemental) body (*jism*).
>
> As for the category (of knowable things) that are in a substrate which are not self-subsistent and embodied – except by way of being dependent (on

their substrate), members of this category are either necessary concomitants of their substrate, or they are not (necessary concomitants). Or rather, that is how it seems to ordinary vision, since in the fact of the matter nothing that does not subsist by itself (i.e. everything but the Creator) actually continues (in existence) for more than the instant of its existence; it may either be followed by (new creations that are) similar (*amthâl*), or by that which is not similar to it. As for what is followed by (new creations that are) similar, that is what are imagined to be the 'necessary concomitants' of a thing, like the yellowness of gold or the blackness of ebony. As for (those characteristics) which are not followed by similars, they are called 'accidents' (*al-'arâd*), while the necessary concomitant is called an (inherent) attribute (*sifa*).

So the knowable things that have actual existence are not more than those we have mentioned.

Now you must know that the world is one in substance and many in form (appearance). So since it is one in substance, it does not transmute (from one thing into another entirely different one: *lâ yastahîl*). And also the form itself is not transmuted, since otherwise this would lead to 'reversing the realities' (*qalb al-haqâ'iq*) – for heat may not (at the same time) be coldness, dryness may not be wetness, whiteness may not be blackness, and the triangle may not be square. But something that is hot can come to exist as cold, though *not* at the same time when it is hot; and also what is cold can come to exist as hot, but not in the same time when it is cold. Likewise what is white may become black, and the triangle may become a square.

So there is no transmutation (*lâ istihâla*), but the earth, water, air, the (celestial) orbs (*al-aflâk*) and all the generated existents (of the sublunar world: *al-muwalladât*) are (only) forms in the (Single) Monad. So (certain) forms are bestowed upon it and that (process of bestowing forms) is called, with respect to their specific shape (*hay'a*) 'generation' (*kawn*). Or (certain) forms are taken off of so that a (particular) name (i.e. attribute or property) is removed, and that is (called) 'corruption' (*fasâd*). So in fact there is no transmutation, in the sense that the actual entity of a thing changes into another (entirely different) actual entity, but it is only (by an entirely new re-creation) as we have explained.

So the world is continually being generated and corrupted (destroyed) at every single instant of time (*zamân fard*). And there would be no persistence for the actual entity of the substance (Monad) of the world, were it not for its receptivity to this 'creative formation' (*takwîn*) in itself. So the world is always continually in need (*faqr*: of the divine creative force). As for the forms, they are in need (of Allah's creation) in order to come out from nonexistence into existence. And as for the Monad, it (is in need) of preserving its existence through that (creative Act), because its existence is unavoidably conditioned upon the existence of the creative formation of that (i.e. the infinite forms) for which it is a substrate.

Likewise (with the dependency on the Creator) of the (purely spiritual) self-subsistent contingent (existent) that is not embodied: it is (still) the substrate for the spiritual attributes and perceptions that it supports, so that its own individual reality may not continue without them. But those spiritual attributes and perceptions are continually renewed in that (spiritual existent) just like the accidents (forms) are continually renewed in the bodies.

In the same way the contingent that exists by itself and is not embodied is the substrate of what it carries of spiritual descriptions and perceptions (that is the bearer of meanings) that its essence may not remain without them. And they are renewed on it just like the renewal (i.e. re-creation) of the forms in the bodies; the image of the body is a form in the monad but the terms (*hudûd*: by which the object is described) are related to the images (of the body, not to the body itself). So the images are the ones which are termed (*mahdûda*), and one of these terms is the monad in which these images appear. That is why they (the philosophers) call the images monad(s) because they take the monad in the term of the image.

This is possibly the main barrier that prevents us from witnessing the reality of the world; we always try to assign an image to every concept, and then we discuss the terms of this image such as its shape, colour, size . . . etc. The Single Monad, however, may not be captured into image, not to mention the Real Himself. Therefore, approaching these matters in any way other than the path of (experiential) divine unveiling will not lead one to the truth of the matter as it really is. No wonder that they (those who rely on their own unaided theorizing) never cease to be in disagreement (about this). That is why the group of the blessed, who are supported by the Holy Spirit, turned to purifying themselves from their own thinking, and to liberating themselves from the bonds of their (natural, animal) forces, so that they became connected with the Greatest Light and saw for themselves (the reality of) the matter as it really is in itself! Because the Real – may He be cherished and glorified – *is* their vision [*Kanz*: 21327], so all what they see is the Real (Chittick 2002: 116–124). As the righteous one (Abû Bakr al-Siddîq) said: 'I have seen nothing but I have seen Allah before it.' So he sees the Real, then he sees His effect in the world; that is to witness how the world emerged as if he witnessed the possible things in their determination state when (Allah) threw what He has thrown on them from His greatest light so they became described by existence after they were described by non-existence [*Kanz*: 548, 1314]. So this (person), who has got this state, the veil of blindness and misleading has been removed for him: '*now We removed thy veil, and sharp is thy sight this day!*' (50:22), '*Lo! therein verily is a reminder for him who hath a heart*' (not only an intellect, see also section 5.1), '*or gives an ear and he is witnessing (the truth)*' (50:37). So (Allah) made knowledge available in witnessing, because the judge judges based on his best guess while the witness witnesses with knowledge not by guessing.

[II.454.1]

According to this passage, Figure 6.1 summarizes the different types of things in existence.

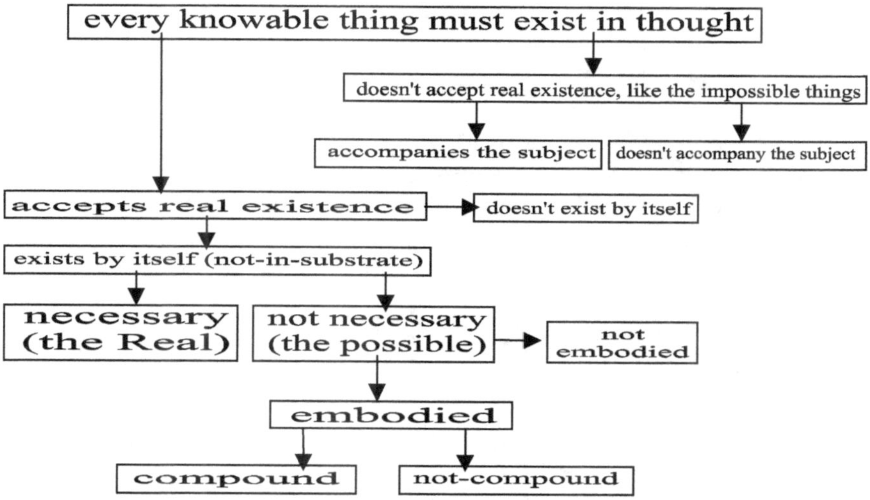

Figure 6.1 Summary of the different types of knowable things.

Also in his *Inshâ' al-Dawâ'ir* (*Constructing the Circles*), Ibn 'Arabî explains the different categories or types of existence in the same way, which he represents schematically in Figure 6.2.

6.5 The greatest element

As we have seen above, although for Ibn 'Arabî the monad or atom (*jawhar fard*) is an indivisible physical unit, it is understood to be composed of even more elementary constituents. This means that there are smaller – but not necessarily physical – 'constituents' that somehow underlie and help manifest the atoms or monads, even though the monad itself is not physically divisible into those metaphysical constituents, but can only exist in manifestation as a substance created through those underlying constituents. So what are these ultimate constituents of the Single Monad?

In his book *'Uqlat al-Mustawfiz* (*The Bolt for the Restless*), Ibn 'Arabî speaks about the 'Greatest Element' (*al-'unsûr al-a'zam*) from which Allah has created the 'Absolute Unseen' which may not be disclosed to any creature, and he indicates there that the creation or 'origination' of this Greatest Element is all at once, without any intermediate or associated causes, as we have seen in section 2.16 (see also *'Uqlat al-Mustawfiz*: 38). So this original, metaphysical 'Greatest Element' that is in some mysterious way the substrate of all subsequent manifest creation – whether purely spiritual, imaginal or physical – is the only thing that in some way underlies, constitutes, or gives rise to the physical monads. The individual monad or atom, however, remains the basic indivisible structure in the physically manifest world.

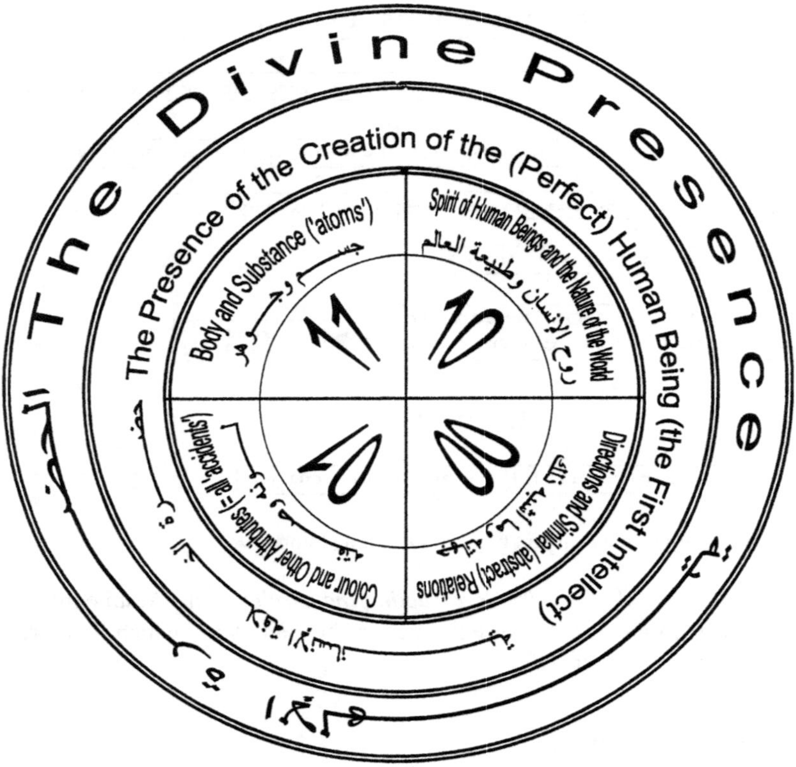

Figure 6.2 The Different Divisions of Existence (source: from *Inshâ' al-Dawâ'ir*, page 19).

Note
The numbers 0 and 1 in the inside circle describe whether the corresponding division is localized (1) or not localized (0); and (second number) whether it is self-subsistent (1) or not self-subsistent (0). The spirit is self-subsistent but not localized, while colour is localized and not self-subsistent.

In the *'Uqlat al-Mustawfiz*, Ibn 'Arabî also mentions that there are 46,656,000 subtle luminous links (*raqâ'iq nûraniyya*) between the First Intellect and the Greatest Element that is their origin (*'Uqlat al-Mustawfiz*: 40). That number is in fact the cubic power of 360 ($360^3 = 46{,}656{,}000$), which is no doubt symbolically associated with the traditional division of the circle (and the divine year, see section 3.2) into 360 degrees and the historical sexagenary system attributed to the Babylonians (see also section 4.7).

Now the resulting relation between the manifest world, the Single Monad (First Intellect) and the Greatest Element can be conceived by analogy to the relation between a building, the bricks and the clay: i.e. the building is made up of similar unit bricks, but the brick itself is made from fine clay. Ibn 'Arabî also says:

> the noble Greatest Element is in relation to the sphere of the world like the (indivisible) point, and the Pen is like its circumference, while the Tablet (i.e. the world-Soul) is what is in between (the point and the circumference). So just as the point meets the circumference with its (whole) entity, so does this Greatest Element meet with its (whole) entity all the aspects of the Intellect, which are the subtle links (the 46,665,000 *raqâ'iq*) that we mentioned before. They are unique 'one' in the Greatest Element, but in the Intellect they become multiple and manifold, because of the manifold receptivity (of the Intellect for knowledge) from the Greatest Element. So there is (only) one 'close attention' (*iltifâta*) for the (Greatest) Element, but there are many faces of receptivity for the Intellect, that is why this (Greatest) Element is more realized in the unity of Its Creator.[9]

Ibn 'Arabî affirms that this Greatest Element is the most perfect thing in existence and that everything other than Allah (including, as we can see, the First Intellect) is somehow derived from it. However, he does not give much information about It, and he even says that he would explain the reality of this Element if he was not sworn not to disclose it. However, does explain, as we have just seen, that this Greatest Element has a special attention (*iltifâta*, like the divine 'special Face' discussed in section 5.2) to the metaphysical 'world of writing and recording' (*'âlam al-tadwîn wal-tastîr*), when the (manifest, including physical) world was still not yet existing in reality (but only in Allah's foreknowledge), and that Allah created the First Intellect (that is the Single Monad) through this special attention (*'Uqlat al-Mustawfiz*: 39).

In the summary cosmological chapter 60 of the *Futûhât*, Ibn 'Arabî alludes more symbolically, in a metaphysical exegesis of Qur'an 68:1 ('*Nûn and the Pen, and what they are recording* ...'), to the 'Greatest Element' when he speaks of the mysterious figure of '*Nûn*' whom Allah appointed as the divine 'chamberlain' (*al-hâjib*) and gave all His Knowledge of His creation, so that Allah – with regard to His Name 'the All-Knower' – never hides from the *Nûn*. And Allah appointed another angel, the Pen – who is the Single Monad/First Intellect/Perfect Human Being – as the 'scribe' for the *Nûn*, 'writing out' all of the divine Knowledge of His creation [I.294.33].

6.6 Analogies in the macrocosms

For Ibn 'Arabî, of course, the mysterious metaphysical relations between the ultimate macrocosmic constituents of creation are repeatedly mirrored in many different 'microcosmic' dimensions of our own life. In each of these different symbolic domains, the initial creation of this higher world and the relation between its elements, such as the First Intellect and the Greatest Element, is subsequently reflected on many different lower planes of existence.

6.6.1 The Black Stone and the Kaaba

The most visible example of Ibn 'Arabî's development of this cosmological symbolism in the *Futûhât* involves the Kaaba, the 'house of Allah' to which millions of Muslims now go on pilgrimage every year. For Ibn 'Arabî, those circumambulating the Kaaba are mirroring the circles of higher angels surrounding the divine Throne [I.50.30]. In that symbolic context, the angels also represent the determining forces of the universal Nature (*Al-Durratu Al-Baydâ'*: 138; we shall come back to this later in section 7.10) and the four Archangels who carry the Throne of Allah are the four main sustaining forces of that Nature.

This centrality of the symbolism of the Kaaba is of course rooted in the fact that Ibn 'Arabî started the first chapter of his *Futûhât* by mentioning his encounter with the Spirit from whom he took all that he wrote in this book, a Spirit whom he met while circumambulating the Kaaba. There Ibn 'Arabî establishes a symbolic correlation the seven circles of *tawâf* that the pilgrim is obliged to perform around the Kaaba during the pilgrimage and the seven main divine Names, in the manner already mentioned in Chapter 3: i.e. each one of these seven Names is responsible for one specific Day of the divine Week of creation. Then he says that his Lord told him: 'the Kaaba, in relation to the all-encompassing (divine) Throne, is like your heart with relation to your body' [I.50.29]. So in fact the Kaaba on the Earth is symbolically like the Single Monad in the cosmos. This analogy also applies to many related details, because the cubic shape of the Kaaba is in fact the simplest structure which constitutes a body that occupies the three spatial dimensions. As Ibn 'Arabî mentioned [III.276.4; see explanations in section 5.4 above], the body is composed of at least eight points, corresponding to the corners of the cube.

But more importantly for Ibn 'Arabî, one corner of the Kaaba holds the mysterious 'Black Stone' (*al-hajar al-aswad*) which, according to tradition, the angel Gabriel brought down from Paradise and gave to Abraham to put it in that corner of the Kaaba. For Ibn 'Arabî, this Black Stone symbolically represents the foundational role in the process of creation or manifestation of the 'Greatest Element'. In other words, circumambulating the Kaaba starts from the south-eastern corner in which this Black Stone resides, and the pilgrim is supposed to make seven rounds (counterclockwise) around the Kaaba: this corresponds symbolically to the way the Greatest Element first gives rise to or communicates to the Single Monad or First Intellect, after which the Intellect brings forth the world of manifest creation in the seven divine Days. According to tradition, the Prophet Muhammad said that this Black Stone resembles 'Allah's right hand on Earth' [*Kanz*: 34729]. As is well known, Ibn 'Arabî holds that the 'Universal Reality' – which is also another name for the Greatest Element, because it is the origin of the Single Monad [I.119.10] – is identical to the Spirit of Muhammad himself, as that Spirit is also, according to a number of widely known hadith, 'the first thing to be created' [*Kanz*: 31917].[10]

Thus, at the very beginning of the opening chapter of the *Futûhât*, when Ibn 'Arabî begins to speak about the underlying metaphysical reality – i.e. the

'Greatest Element' and first creation – symbolized by the Black Stone that resembles Allah's right hand, he says in poetry:

> *People are ignorant of its Essence, so some say it is dense, while others say it is subtle.*
> *He (the Spirit) said to me, when I asked why they do not know It:*
> *'Only the noble may truly know (recognize) the noble!'*
>
> [I.47.22]

Ibn 'Arabî then proceeds in these opening pages to give many mysterious symbolic details about what Allah creates in the Human Being (i.e. the Single Monad or First Intellect) and in the world with each round of the seven circumambulations around the Kaaba, and he relates that metaphysical teaching to the seven main Attributes of Allah which are responsible for the seven Days of the divine creative Week [I.49.32].

As shown in Figure 6.4, just the Greatest Element makes the Single Monad which scans the states of the world in seven Days, the pilgrim in Hajj has to make seven rounds around the Kaaba anti-clockwise. This circumambulation starts from the eastern corner where the Black Stone resides and moves towards the Shami corner. This clearly supports the analogy between the Greatest

Figure 6.3 The Kaaba, with people on the Hajj circumambulating it. The Black Stone appears in the front corner, near the door.

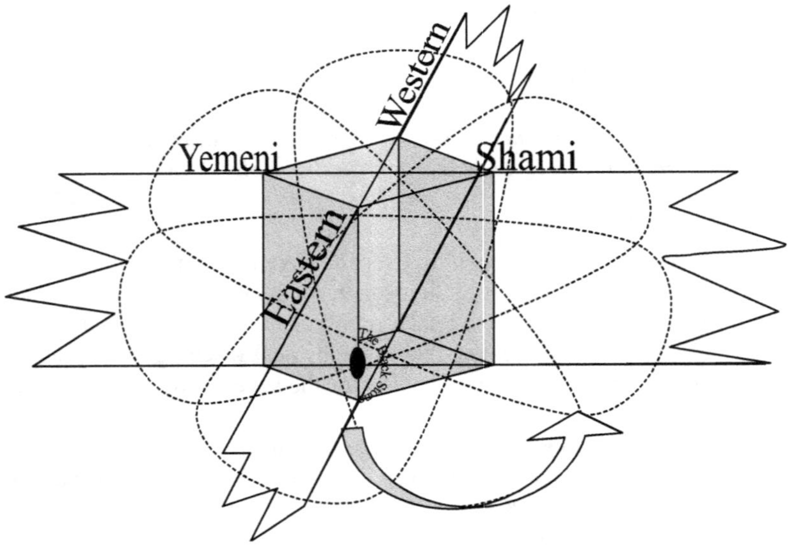

Figure 6.4 How circumambulating the Kaaba is similar to the Greatest Element's creation of the Single Monad. Circumabulation starts – anticlockwise – from the eastern corner where the Black Stone is, and that corner is for the Pole.

Element and the mysterious Black Stone, especially, since we have said above that the Black Stone resembles 'Allah's right Hand on Earth' [*Kanz*: 34729].

6.6.2 The spiritual hierarchy

Another important example of the symbolic analogy between the metaphysical process and figures of the 'macrocosm' and more human realities is the hierarchy of the spirits of the prophets and saints (*awliya*) – a central theme that runs throughout the *Futûhât*. To summarize the cosmological aspect of that theme, Ibn 'Arabî presents the lower realms of the cosmos as being ruled by a complex spiritual hierarchy – largely 'invisible' to most human beings, though not to the spiritual 'knowers' – consisting, among others, of the spiritual 'Pole' (*al-qutb*), the two Imams, the four Pillars (*awtâd*), the seven Substitutes (*abdâl*), the eight Nobles (*nujabâ'*), the 12 Chiefs (*nuqabâ'*), in addition to other lesser-known groups.[11] Each of these groups and figures has a special corresponding spiritual responsibility, some of which are have to do with maintaining the wider cosmic order. Those pure spirits have also have an ongoing series of living human 'representatives' or 'agents' (s. *nâ'ib*) amongst us [II.6.6].

Some of these members of the celestial spiritual hierarchy, as Ibn 'Arabî presents them in scattered passages of the *Futûhât*, are assigned cosmological functions symbolically related, for example, to the 12 zodiacal signs, the seven heavens, and the seven 'climes' or geographical regions, the four cardinal points, or even the four corners of the Kaaba [II.13]. In particular, the highest level of the 'Pole', in this hierarchy, is the figure who apparently corresponds to the lofty metaphysical position of the Single Monad – which Ibn 'Arabî often pointedly refers to simply as 'the Reality of Muhammad' (*Al-Mu'jam Al-Sûfi*: 158).

6.6.3 The world of letters

Also yet another important symbolic analogy between the metaphysical macrocosm and more familiar human realities, which again runs throughout the *Futûhât*, is the 'world of letters'. Ibn 'Arabî considers the letters of the Arabic alphabet – given their central place in the culminating divine revelation of the Qur'an – to constitute in themselves 'a real world like us', since they are servants of Allah just like ourselves [I.58.13]. He begins his detailed explanation of their symbolic metaphysical and cosmological functions in a long section in the opening chapter of the *Futûhât*. As he explains there, the Arabic letters also have symbolic hierarchy similar to the spiritual hierarchy of the prophets and the saints. Thus they also have Pole, which is the letter *alif* (ا), the first letter of Arabic alphabet; two *imâms* (Leaders), which are the two other vowel letters *wâw* (و) and *yâ'*; four *awtâd* (Pillars), which are the letters *alif* (ا), *wâw* (و), *yâ'* (ي) and *nûn* (ن), that together provide the essential Arabic grammatical indications (*'alâmat al-i'râb*); and seven *abdâl* (Substitutes), which are the letters *alif* (ا), *wâw* (و), *yâ'* (ي), *nûn* (ن), and the three key pronoun markers *tâ'* (ت), *kâf* (ك) and *hâ'* (ـه)] [I.78.18]. But the important facts about the metaphysical relations between the world, the Single Monad and the Greatest Element are particularly clearly developed in Ibn 'Arabî's teaching here regarding the metaphysical dimension of this world of letters. Thus he clearly states that:

> For the totality of the letters (like the world) may be deconstructed into the *alif* (corresponding to the Pole and the Single Monad) and put together from it (the letter or sound) *alif*, but it can not be deconstructed into (any of) them. However it too can be deconstructed – in our symbolic estimation – into its spiritual principle (*rûhâniyya*), which is the (primordial) 'Point' (of Greatest Element) – although (in fact) the one can not be (further) deconstructed.[12]
>
> [I.78.23]

Here again the 'Greatest Element' is assumed to be the underlying principle or substrate of all creation and manifestation.

As we have mentioned on several earlier occasions, Ibn 'Arabî often attributes his source for some of the most important metaphysical knowledge about the world to the spiritual Pole Idrîs ('*mudâwi al-kulûm*'). In describing the relation between the originary 'Point' of the 'Greatest Element' and all the manifest cosmos, in chapter 15 of the *Futûhât*, this Pole tells him:

> The world (physical cosmos) exists between the circumference and the point (of the Earth at its centre), arranged according to the levels (of its orbs) and

> the smallness and greatness of the orbs. And (he said) that the sphere that is closer to the circumference is wider than that which is inside, so its day is longer, its space is larger, its tongue is more fluent and it is closer to realizing strength and purity. And what goes down to the (material, earthly) elements is less than this (high) level, on down to the sphere of the Earth. But each part in every circumference matches what is below it and what is above it with its (whole) entity: no one is greater than the other, despite the fact that one is larger and one is smaller! . . . And all match the point with their entities – and that Point, despite its smallness, matches the parts of the circumference with its (whole) essential reality (*'ayn*).
>
> [I.154.22]

It is worth mentioning here, as a modern analogy to this earlier symbolism, that one of the most compelling consequences of Quantum Mechanics is that everything in the physical world of 'particles' can also be expressed as waves that have different wavelengths. Electrons for example – though they are particles – have a wavelength. Even the Earth has a distinctive wavelength. When the mass of the body or particle becomes larger its wavelength becomes smaller. So the wavelength becomes even larger for massless particles such as photons (light).

6.6.4 The hierarchy of divine names

We have already discussed the unique Unity of Allah and the diversity of His divine Names in section 5.3. In Ibn 'Arabî's wider metaphysical perspective, the divine Names – just like the spiritual world of the angels, prophets and saints – are also arranged in a corresponding specific metaphysical hierarchy. In section 3.1, we have already encountered the four and seven fundamental divine Names – which correspond to the four *awtâd* and the seven *abdâl* respectively. But Ibn 'Arabî talks about this hierarchy of the divine Names in more detail early on in chapter 4 of the *Futûhât*. Here we want only to draw the attention to his explanation there of the difference between the two key divine Names 'Allah' and 'the All-Merciful' (*al-Rahmân*).

Already in his *Al-Tadbîrât Al-Ilâhiyya*, Ibn 'Arabî compares the relation between those two Names with the cosmological or metaphysical relation between the Throne and the Single Monad, as we explained in section 4.6 above (see also *Tadbîrât*: 89). Because the Single Monad is itself also called 'the Throne', Ibn 'Arabî says that – in cosmological language – Allah mounted on the Throne of the Single Monad, while '*the All-Merciful mounted on the Throne*' (following the literal Qur'anic verses that we discussed earlier in section 1.4. So the relation between these two Names is like the relation between the two divine 'Thrones'. It is to be noted here that the Name 'the All-Merciful' is more specifically described by all the multiplicity of the divine Names, just like the Name Allah. For as Allah said in the Qur'an: '*Call upon Allah or call upon the All-Merciful, Whoever you call upon, to Him belong the most beautiful Names*' (17:110). So the Name Allah is also described by all the divine Names, but the

Name 'the All-Merciful' is closer to the multiplicity of the world through those diverse divine Names. Likewise, the Throne that encompasses the heavens and Earth is a place of manifest multiplicity, whereas the Single Monad, in its mysterious intrinsic 'hiddenness', is more described by the transcendent divine qualities of unity and oneness.

6.7 Spreading the shadows

In discussing the divine Name 'the Subtle' (*al-latîf*),[13] Ibn 'Arabî says that there is great *subtle* knowledge in 'the withdrawing of the shadow and its spreading' (alluding to 25:45–46), and that is why Allah made that symbolic image a guide to Him [IV.238.2]. As we pointed out before, Ibn 'Arabî views the world as a structure made 'according to the image or form' of the Real (see section 3.1). In the same way, he considers the manifest world as the 'shadow' of the Real, or of the Universal Intellect (and ultimately the 'Greatest Element'). He bases these arguments on the following two verses from the Qur'an: '*Hast thou not seen how thy Lord hath spread the shadow – and if He wished, He could have made it still? Then We have made the Sun a guide unto it. Then We withdrew it into Us, with a diminutive withdrawal*' (25:45–46), and we have already drawn attention to this basic relation between the divine Image and its 'shadow' in section 3.1. In many places [III.12.3, III.106.7, III.281.32] Ibn 'Arabî emphasizes that the shadow of anything is on its own image; therefore:

> Know that the Human Being (*Insân*),[14] since he is the 'likeness' of the divine Form, is like the shadow of a person, which does not ever leave him, although sometimes it appears to the senses and sometimes it is hidden. So when the shadow is hidden, it is still implicit (*ma'qûl*: 'present-to-the-intellect') in that person; but when it is manifest, then it is visible to the eyesight – (at least) for whoever (actually) sees it. So (likewise) the Perfect Human Being (*al-insân al-kâmil*) is implicit in the Real (*al-Haqq*), like the shadow when it is hidden by direct sunlight so that it does not appear. For (the Perfect) Human Being always continues to be, eternally and perpetually. That is why (the Perfect Human Being) is always witnessed by the Real, since He is described (in the Qur'an) as having Sight (*al-basîr*: 17:1, 40:20, etc.).
>
> Therefore when '*He spread* His *shadow*', it appeared according to His Form: '*Hast thou not seen how thy Lord hath spread the shadow – and if He wished, He could have made it still?*' (25:45) – i.e. stably remaining with Whom is His shadow. But if He does not spread His Shadow, no individual reality (*'ayn*) of It will be manifest in sensible existence, except to Allah alone (in His Foreknowledge). So His 'Shadow' (i.e. the Perfect Human Being/Single Monad) has always been with Allah and will always be with Allah, for he continues Allah's continuing, while everything other than the Perfect Human Being (Single Monad) only continues (in existence) through Allah's causing them to continue.
>
> [III.187.15; see also I.458.33 and II.607.14]

Therefore, the Single Monad or Perfect Human Being is the primordial Shadow of the Real, and the manifest world is in turn the shadow(s) of the Single Monad. The world appears in existence, after being hidden in the 'Unseen' (of the divine Foreknowledge), through Allah's 'spreading the shadow' of the Single Monad in the eternally renewed act of Creation. For if God so wished, He could have left His Shadow unmanifest, which means that the created world would have stayed as it was in Allah's eternal Knowledge, instead of coming out into real existence spread over space and time.

However, the 'shadow' of the Single Monad is one like him: that is why Allah withdrew his shadow 'with a diminutive withdrawal' – like the withdrawal of the shadow that normally occurs at sunset (see I.459.5) – so that it disappears in order to appear again in a different outward form, just as the sunlight disappears in order to eventually make another day. So at every 'Day' (instant) the Single Monad makes a new shadow in all creation. So the world is the ongoing collection of those instantaneous shadows, as explained in sections 3.6 and 4.3.

When the shadow is spread (before and after noon) it is itself a guide to the Sun, just as the Sun was an actual guide to it by its effect which caused the shadow itself [I.459.9 and also IV 238.29]. Likewise our shadow-like existence (our actual individual entity) should guide us to realize the Real manifested in ourselves – being His shadows – by looking at His shadow (i.e. the Single Monad or Perfect Human Being) who makes manifest the shadow of our creation – and by proceeding then from the Perfect Human Being to the Real Himself, the divine 'Light' or Sun Who ultimately caused this shadow (that is our existence). The seeker of the Real (*murîd*), therefore, keeps seeking the Real in these shadows and in the Shadow (of the Perfect Human Being) that is their Source, until he comes to realize that Source more fully and directly. So when he achieves that high state of realization, he himself – being one of these shadows – will be extinguished or 'drowned' (n. *fanâ'*) or united (n. *ittihâd*) with the Real – because the shadow can not see itself and the Real at the same time – except perhaps for an instant of time, comparable to the case of noontime at the equator, where the shadow of the object disappears for a single moment.

But Ibn 'Arabî always clearly points out that after the 'enlightened' shadow-subject witnesses the Real, and continues to watch Him, then he will realize that he is as he was before and after his direct witnessing – except that before that witnessing he did not realize that what he was witnessing was the Real – just as the Sun is the Sun before and after the noontime disappearance of the shadow.[15] However, in fact the shadow at noon hides in the object and not in the Sun, so in fact the realized knower indeed does not witness the Real Himself (the Essence), but witnesses only the 'real-through-whom-creation-takes-place' (*al-haqq al-makhlûq bihi*), which is the Perfect Human Being (the Single Monad, or at most the Greatest Element). So the Knower may at most realize his or her essential unity with *this* reality, which is only the higher 'Shadow' of the Real.

It is here, Ibn 'Arabî implies, that many of the Sufis' ecstatic expressions (or poetic metaphors) have sometimes been misunderstood. For example, Ibn al-Fârid in his famous poem, the *Tâ'iyya*, speaks about 'unification' (*ittihâd*) in various forms (Mahmûd 1995: 241, 248, 252 and 276). Of course Ibn 'Arabî himself has sometimes been accused of believing in *ittihâd* and *hulûl*, but he was always very careful not to mention these words and directly denied such doctrines [III.298.30, IV.81.23].[16] That is why in his prayers Ibn 'Arabî asks to unite with the Prophet Muhammad's spiritual essence (*dhât*), who is the Perfect Human Being. For example he says in his book of prayers, *Al-Salawât Al-Faydiyya*:

> O God, pray for him (Prophet Muhammad) a prayer by which my branch may connect with my root, and my part with my whole, until my essence is unified with his essence and my attributes with his attributes, and the eye is satisfied with the eye, and disunity flees away from disunity.[17]

In yet another of his prayers, he very clearly expresses what we have just observed in his remarks about the divine 'shadow':

> Oh God, no god but You, it is You Whom we worship and You Whom we witness; ever returning to You, nothing but You. I ask of You, by You, You Yourself, Yourself, Yourself, You Who (have) no 'he' other than 'Hû' (the divine Essence), (I ask of you) to withdraw from me the shadow of (my natural bodily) formation, so that I may witness my (true) self bare from any description that is a veil which prevents me from witnessing You as 'I',[18] and purify me from any attribute or influence that makes me see any (purely personal) share (in existence): 'for *everything is perishable but its* [*or 'His'*] face'[19] (28:88), '*but to Allah all things are returned*' (42:53). Oh my God, pray for Your messenger, our master Muhammad, who is apportioned with this perfect abolishment (*mahw*), and complete integration (*jam'*) that is beyond perfect wisdom.[20]

For Ibn 'Arabî, these symbolic analogies between the shadows and creation are indeed very important existentially because Allah made them as guides for us, in all the ways we have just reviewed. But what is particularly important here, in terms of Ibn 'Arabî's actual cosmology and understanding of time, is that the created entities of the world that are the evanescent 'shadows' of the Single Monad are continuously re-created, by one instantaneous act of creation after another, in a process similar to the spreading of the shadows by the Sun in the course of the day. In the following section, we turn to some further illustrations and implications of that cosmic process of creation.

6.8 Creation scenario: the world as a movie

Later Muslim writers, especially the great masters of mystical poetry in Persian and other eastern Islamic languages, developed a great range of familiar symbolic forms intended to elaborate and communicate the basic Qur'anic imagery for the cosmological processes and their symbolic expression that we have discovered in the writings of Ibn 'Arabî. One of the most powerful and multi-faceted of those images was Ibn 'Arabî's famous account – reminiscent of the Sun-Line-Cave section of Plato's *Republic* – of the universe as a vast 'shadow-theatre', a true 'divine Comedy' (Morris 1993: 50–69). Thus in contemporary terms, based on the concept of perpetual cosmic re-creation that we have explained above, we may envisage the world as like a movie being displayed on a computer monitor. It is quite fascinating to discover that this analogy is quite accurate in most of the details, even regarding what happens inside the computer, since the Universal Intellect/Perfect Human Being/Pen can be considered as a kind of 'supercomputer' which creates, organizes the world and displays it in the 'Universal Tablet' (of the world-Soul). Ibn 'Arabî has already asserted that the world appears as 'living, hearing, seeing, knowing, willing, able and speaking' with the same seven fundamental Attributes of the Real. This is because the world is His divine work, and as Allah said: '*Say: "everyone works according to his own type"*' (17:84) [II.438.19]. This is also equivalent to what we have repeatedly noted: that the Perfect Human Being and the world (including its human beings) are all created 'according to the Form or Image' of the Real, and that they are also the second-level shadows of the Real (see for example [I.163.20, II.652.25, III.343.25], and also [*Kanz*: 1142, 1148]).

Likewise the computer today is certainly created as a certain kind of 'image' of some specific aspects of the human mind, and certainly the way the computer works resembles the human mind in many respects. However the shadow or the image in the mirror (or photocopier) resembles at best only one facet of the original. Likewise human beings, for example, do not fully resemble the Real, just as computers do not (and cannot) fully resemble humans in many other respects.

As we have seen, Ibn 'Arabî repeatedly showed that in many respects both the world and the essential (spiritual) human being work in the same way. That is why he generally considers the world as a great human being (*al-insân al-kabîr*) and the human being as microcosm (*'âlam saghîr*) [III.11.18]. Just as our own world is essentially constituted by the meanings, images and states that are reflected in our spirit, soul and intellect, so the world also reflects the divine 'meanings' brought into creation by the Universal Intellect, so that the phenomena that we perceive are the doubly reflected forms of these original meanings. However, Ibn 'Arabî also repeatedly asserts that the world would not exist if the observing 'eye' or 'I' of the viewer was not also always there.[21] So because of the underlying reality of the created world as essentially 'imaginal' forms or multiple 'reflections' or shadows, we can only understand the cosmos if we understand how we perceive it, since for Ibn 'Arabî it is all ultimately, and quite literally 'in the mind' – albeit a different kind of 'Mind' at each level of manifestation.

It is known that the movie which is displayed on the cinema screen is composed of a large number of succeeding still pictures that pass rapidly before the eye at very short intervals, so that the human mind observes only smooth changes between those rapidly successive pictures. By running this movie at the proper speed we feel (by illusion) as if a normal motion of objects and images is happening on the screen. So if we suppose that the screen has no visible edges, and especially with the new technology of three-dimensional (laser holography) movies, it would be very hard initially to distinguish this illusion from reality.

Now if we examine how the picture is displayed on the screen of the computer monitor, we realize that it is even more closely similar to Ibn 'Arabî's view of creation. In this way the whole cosmos is a combination and rapid succession of imaginal forms (images, reflections or 'shadows') that are created by or through the Single Monad in a similar manner to the single electron beam which is creating the picture on the computer screen one pixel at a time.

As Ibn 'Arabî has pointed out, the Single Monad is continuously and perpetually 'wearing' new forms which make us see and otherwise experience motion. When we open our eyes we see a picture of many things around us, and if we keep on watching we see things moving. Each mental picture is also created in series and not all at once, as we have noted before (sections 5.5 and 5.6). In every single instance and at each single point of space there exists a monad with a specific unique form, and we have explained before (sections 3.15 and 4.1) that it takes the Monad a full divine 'Week' of creation to appear in this form, but this divine Week for us is like a single instant. This same monad, still in the same instant of time – for us, but a full divine 'Week' for the Monad itself, since we only exist for one moment in this Week: see section 3.6 – takes another form but in another point of space, and so on. So in one single instant the picture that we see is a combination of a huge number of reflected forms of the same Single Monad. He or It scans the whole of space at no time (for us) and without real motion (on the part of the Single Monad), because space itself is what we subjectively experience as a consequence of the succession of forms within this monad, and motion is meaningless when we talk about one single all-encompassing entity. It takes the monad a full 'Week' of creation (i.e. seven 'Days of event': one for each direction of space – up, down, right, left, front and back, and one for the observer – time) to scan all the states in the cosmos, but since each one of us is one single state – as observers, not as bodies – we live a single moment in each full 'Week of event', in which we observe the other states around and within us as the traces or memory of the forms left over by the Monad after it has created those states. Ibn 'Arabî succinctly referred to this cosmological fact, in a favourite image of the later mystical poets, right in his Forward (*khutba*) to the *Futûhât*, when he said:

> Then He released the Breath, so the water waved because of its vibration and foamed . . . Then the water diffidently withdrew and returned back heading for the middle, and it left over its foam on the shore that it produced. So it (the world) is the churning of this water that contains most things.
>
> [I.4.7]

The 'water' here refers to the Single Monad itself (or the Greatest Element: *Al-Mu'jam Al-Sûfi*: 812–817, 826–828) because (in the famous expression of the Qur'an) 'every thing was created from the Water', and the 'foam' is the created forms (or their images) left over by the Single Monad after it has 'scanned' into existence the created world (in six divine Days) and then returned back to the middle to start over a new picture [II.438.3] on Saturday. That is why Ibn ʻArabî also affirms that the cosmological ʻruler' of the last Day of time (Saturday) has the ability of holding and fixing (i.e. memory), in order to hold the cosmic picture and integrate it with what follows (see also section 3.5, Saturday as the Day of Eternity).

When this perpetual creative process is conceptually ʻstopped' and taken in isolation, all this will form a kind of ʻstill picture' of things around us, including ourselves both as bodies (matter) and as spirits or states of realization (meanings). Within this conception, the dynamic manifest world, then, is the instantaneous, continuously renewed succession of these slightly changing still pictures. As we showed in section 2.6, motion is observed because things successively appear in different places, but indeed there is no actual motion: for the observed objects are always at rest in the different positions that they appear in. Allah is constantly re-creating the cosmos in ever-renewed forms.

Now what we have just mentioned is exactly like what happens on the screen of the computer monitor: when we look at the screen at any instant of time, we see a still picture that is composed of an array of dots (pixels) in the two dimensions of the plane of the screen (for example 800 horizontal by 600 vertical pixels). This still picture is made by a single electron beam that scans the screen over and over again, one pixel at a time. It starts from the bottom left corner of the screen and scans horizontally all the 800 pixels (one line), then it switches back to the left to make the second line, and so on till all the screen is scanned, ending up by the upper right corner; then it switches back to start a new picture from the bottom left corner again in the same way. Because this process is performed at very high speed or refresh rate (around 100 million times per second), we only see a continuous picture in the two dimensions; we never see the pixels being drawn one by one. By watching the succession of pictures, we observe motion. While the beam creates them, each pixel on the screen wears a specific form of a different colour and intensity that (slightly) changes from one still picture and instant to the other. This momentary form that the pixels wear every time they are scanned lasts only during the very short time that the beam is in its place. Once the beam leaves the pixel for the next one, the form vanishes intrinsically; we only see the traces of these forms for a short time till they are scanned again to wear a new form.

In terms of Ibn ʻArabî's understanding of the cosmogonic process, the electron beam here is like the Single Monad/Pen/Intellect; the screen is like both our imagination and, outwardly or objectively, the effective ʻsubstrate' of creation that Ibn ʻArabî calls ʻthe Dust': *al-ʻamâ'*); while the cosmos is like the series of pictures on the screen, which are printed on our imagination and in the Universal Tablet. Ibn ʻArabî's description of the world is identical to this example of the computer monitor, even the names that he gives to the Single Monad as the ʻHigher Pen' and to the cosmic Soul as the ʻHigher Tablet', indicate that the process of creation is similar to the process of a pen's writing on a tablet, which is also similar to the electron-beam writing on the screen. To take yet another of Ibn ʻArabî's most favoured cosmological images, we creatures are the ʻletters' and the 'words' that are spoken by the Creator (through the creative, existentiating 'Breath of the All-Merciful'), after having been written down by the Higher Pen on the Higher Tablet (see section 7.8).

With a closer examination of this example, we can now realize the meaning of the intertwined days and the taken-out days, and also the significance of the (normal) week and Saturday – as the Day of Eternity – that we explained in Chapters 3 and 4. If an observer was in one point of the computer screen (as we are individually in one specific place in the world), a week for him or her would be a combination of 800 by 600 instances of 800 by 600 'Weeks of events', because each point will be present (created) only while the electron beam is on it, so it stays only one part of the 800 by 600 pixels 'created' during each run (original Week/Day). Similarly, each entity in the world lives only one 'instant' (which equals an outwardly observable day over the entirety of entities in the world) in each 'Week' of creation: i.e. only the time when the Single Monad creates it or appears in its form. Therefore, a week for this point would be a combination of different instants of many other Weeks of the original Weeks of events. Also, because Saturday is the divine Day of manifest creation when each entity realizes itself – and it does not realize itself (or the world) while being created in the other six Days (see section 3.6) – we, as entities living in the world, realize only the Saturdays' instances, those that are manifest in time. Therefore, all our life (as time) is Saturday, while the world as space is the other six Days. In other words: the motion of the electron beam horizontally and vertically on the screen creates the space in which the picture is viewed. This is identical to the cosmic Week-Days of creation from Sunday to Friday, while the motion when the electron beam returns from the right-top to the lower-left corner is identical to Saturday or the manifest instant of time, because time is the motion by which we observe the succession of the pictures.

In section 7.10, we shall talk about dimensions in more detail, but it is good to remember here what we have already said in section 2.11 about the two cycles of life (see in particular Figure 2.1). The screen of the computer monitor is of course two-dimensional (2-D). So the starting of the scanning of the screen from the lower-left corner resembles the starting of the world by creating the Real (0-D) as the pixel itself, and then the angels in 1-D. This in the real world, according to Ibn 'Arabî's account of the astrological cycle of life, takes 11,000 years; but for the case of the computer monitor it takes a very short but also fixed time, in which a fixed number of motions or jumps happen – 800 pixels for example. Then after finishing the first line and by starting the second one, this scanning or creating process moves us into two dimensions, to start the world of jinn, while the 1-D world of angels continues. The scanning of the 2-D screen continues until it is finished after a short but fixed time. There is no third dimension in normal computer monitors, but there is in the outside world: there, according to Ibn 'Arabî, when this manifest stage of creation starts after 54,000 years, it marks the creation of *dunyâ* ('this lower world'). Then the after-world starts in yet another dimension (4-D) – according to Ibn 'Arabî after 9,000 more years, although we have not moved yet into it (see section 7.10).

It is fascinating to notice that this analogy between the world and the computer applies to many of the details – not only with regard to the visible manifest world, but also to the spiritual world that is analogous to the processor, the memory, the hard disk and the software running inside the computer. It is enough to notice that the processor of the computer – no matter how fast it may be and how many jobs it may do very quickly – can do only one elementary job at a time. So from outside we see multiple jobs and multiple images, running and interacting, but from inside only one thing is manipulating all that, one by one, in series, one single bit at a time. For example when we say that the speed of the computer central processing unit is 3.4 GHz, this means that it can manipulate 3.4 billion bits every second, and the bit is the smallest piece of information, which is the digital state of either 1 or 0, yes or no. This is indeed a huge speed, yet surely the speed of the Single Monad creating the world is much larger, as Allah plainly said: '*and He is the fastest of all calculators*' (6:62) – although this is normally interpreted with regard to the Judgement Day.

7 The Single Monad model and its implications for modern physics

This existence that we know in practice,
is not like the existence that we know by unveiling.
The mind does not know it, and thinking does not recognize it,
but remembrance (dhikr) will show it, while the secret will hide it
from the minds that are veiled by habits:
that is why they do not recognize what the secrets[1] *understand.*

(*Dîwân*: 229)

As with most cosmological hypotheses, past and present, testing Ibn 'Arabî's cosmological model and his understanding of time is not easy. Although a preliminary test is simply its ability to resolve some of the standard paradoxes surrounding time, as we shall shortly see below, more credible tests may not be very easy to perform, and incontrovertible tests may be inaccessible – since Ibn 'Arabî frequently reiterates, beginning with the very Introduction to the *Futûhât*, that his most distinctive cosmological insights are ultimately based on forms of inspiration, 'unveiling', and divine knowledge that are beyond the level of the ordinary human intelligence.

So in this final chapter we shall discuss some of the consequences and possible methods to test the Single Monad model that we outlined in Chapter 6. Some persisting paradoxes in current models will be discussed, such as the historical Zeno's paradoxes and the recent EPR paradox (sections 7.4 and 7.6). We shall then go on to discuss some of the apparent consequences of Ibn 'Arabî's cosmological model for various principles of physics, cosmology and philosophy. We have already seen that, in this model, space and time are truly unified, and that there is a drastically different conception of motion. Most other basic properties of matter, like mass, density, softness and transparency, will be also affected, suggesting the eventual possibility of conceiving a kind of 'science' dealing with certain phenomena of mysticism and the supernatural. In addition to that, in this cosmological model, causality itself will have a different sense, which would enable new perspectives for reconciling cosmology and theology.

7.1 Testing possibilities

One of the main consequences – and principles – of this model of cosmogony and creation, as we have seen in the previous chapter, is that the creation is being continuously refreshed, like a kind of computer screen. A first possibility could be, therefore, to measure the 'refresh rate' of creation, which is also the smallest quantum of time, or the length of the moment (*al-zaman al-fard*). As we have noted, this could be expected to be equal to 24 hours (or 24 × 60 × 60 seconds) divided by the whole number of states or entities in the world. With such an extremely small quantity of time, there is no possible device that could measure it.[2] If we think, for example, of using high-speed cameras and then replaying their images in slow motion in order to see the flickering,[3] we shall find that this is not possible even in theory, no matter how fast the camera might be. Since the camera, no matter how fast, is itself part of the world whose re-creation it is supposed to observe, it will not be able to see its state of non-existence because then it will not be existing itself. Similarly, we ordinarily do not notice the re-creation of the world because we are part of it. However, a specially detailed study of the memory and other human internal and external senses would be necessary to judge how the actual process of perception of the world is happening with the re-creation principle in mind, both in our 'normal' state and in those specially enlightened perceptions of the 'knowers' which Ibn 'Arabî points to as the ultimate root of his cosmological thesis of 'perpetual re-creation'.

On the other hand, other promising domains for testing and examining this model would involve the properties of light particles (photons) and other elementary particles. The peculiar properties of photons that are normally treated as electromagnetic waves can be a starting point to test the model above. The photon, being a wave, has a probability of existence anywhere in space and at any time as long as it is still undetected; it takes fixed space-time co-ordinates only after it is detected either by the eye or by a device. We can therefore say that the photon of light, being the fastest in nature, does not undergo any re-creation, and that is why it is the fastest, because creation or re-creation is a process that takes the smallest quantum of time. Therefore one key to test the Single Monad model experimentally could be in the emission and absorption of light by different known processes, or in the process of converting light into mass (and vice versa) through electron–positron annihilation and pair production (see further below).

The best and easiest potential domain of investigation in this regard is to look at the electron orbits around the nucleus, where it is known that the electron jumps between the orbits when it absorbs or emits a photon of light. Because this process is quantized, the electron may not exist between the orbits; it is therefore re-created in the new orbit (see also the following section). We can therefore say that the re-creation time (i.e. the moment) equals the time the electron takes to move between the orbits (i.e. actually re-appears in the new orbit). This time-gap, however, is different from one atom to another, or even from one orbit to another inside the atom, because it depends on the kind of atoms and on temperature, etc. It is not possible here to discuss this subject in detail, but it is certainly a good point which may constitute a new subject of research.

We may also look in the annihilation of particles and anti-particles into light and vice versa ('pair production'). For example, when an electron meets with a positron (each has a mass of 9.1×10^{-31} kg) they are both annihilated into a massless photon (energy) according to the famous equation $E = mc^2$. One electron alone can not annihilate, and one positron alone can not annihilate. But it is possible to convert energy into particles, as when two sufficiently strong photons meet; they convert into an electron (negative charge e^-) and a positron (positive charge e^+). It is worth mentioning here that Ibn 'Arabî has apparently referred to phenomena that have only been known scientifically after the discovery of elementary particles at the beginning of last century. For example, he says:

> When two monads or atoms (*jawharân*) [like two photons] are joined, it is as though they are two bodies. That is to say, when they are joined with each other, each one of them can be called a body (*jism*) so that in this respect they are two bodies, as He said: '*and of every thing, We created two pairs*' (51:49). He actually created one pair – masculine and feminine, for example. But He called it 'two pairs' for this reason that we just mentioned, because each one of them alone without the other is not a pair, but when another is added to it then each one of them may be called pair (*zawj*), so they are two pairs.
>
> [I.721.18]

In other places he says that the body is composed of at least eight points, like a unit cube which is similar in shape to the Kaaba [III.276.6] (see also section 6.6). We have already seen in Chapter 6 that, in Ibn 'Arabî's cosmological symbolism, the Kaaba represents the Single Monad. Therefore, we may look at the possibilities of how mass emerges out of massless photons in annihilation and pair production and see if this phenomenon can be related to the cosmogonic conceptions of the 'Week' of creation explained in Chapter 3 and the 'intertwining of days' explained in Chapter 4.

Yet another related and handier possibility of investigation involves the refraction of light in transparent materials. As is well known, light slows down when moving in transparent materials such as air, water and glass; usually the denser the material, the more slowing light suffers, which is indicated as a higher refraction index. The reason why this is the case is because light is absorbed by atoms along the path and then is emitted again by almost every atom along its path. Any absorption and emission of light can be related to the phenomenon of re-creation, so the re-creation hypothesis can be investigated via the phenomena underlying different refraction indexes. The refraction indexes of hundreds of minerals and their density and other properties are already widely available.[4] So it may be possible, by comparing these data, to find some correlation between the refraction indexes and the underlying refresh rate of re-creation.

Testing this model could also be done by computer simulation, especially since Ibn 'Arabî himself (in *Kitâb Ayyâm Al-Sha'n*: 11–16) invited people who can do more accurate calculations to calculate the intertwining of Days on a smaller scale than hours (for example, minutes and seconds), as we pointed out in section 4.4. This simulation would have to take into account that, as we mentioned in section 3.6, Ibn 'Arabî's theory means that the world is created in six comic 'Days' as space and then displayed on the last Day of creation (Saturday) as time.

In general, the possibilities of testing these hypotheses in physics are diverse and worth trying, but here we are more interested in the theological and philosophical consequences developed in the following sections.

7.2 Timeless motion

To start investigating some of the consequences of the cosmological principle of continual re-creation, we can consider the case of the throne of Bilqîs, which Ibn 'Arabî (and his later commentators) have discussed in his *Fusûs Al-Hikam* and elsewhere. In most accepted understandings, motion is always associated with time, because, no matter how fast objects move, they need a time greater than zero to reach any point other than their original place. For example, it is well known in physics that the photon of light is the fastest particle in the physical world and that it travels at 300,000 km per second, so it can circle around the Earth seven times in one second. However, this huge velocity is limited and not instantaneous, so apparently physical motions always need time.

In the Qur'an, however, we read the story of the Prophet Solomon and Bilqîs, the Queen of Sheba, when Solomon said to his men:

> '*Which of you can bring me her throne before they come to me in submission?"* [Then] *a stalwart of the jinn said: "I will bring it to you before you can rise from your place, and I verily am strong and trusty for such work*'
>
> (27:39)

But even this really quick action takes time, though very little, so

> *said the one who has knowledge of the Book:*[5] *'I will bring it to you before your gaze returns unto you!'* [i.e. in less than the blink of the eye, actually immediately]. *Then when he (Solomon) saw it placed steadily before him, he said: 'This is by the Grace of my Lord to test me whether I am grateful or ungrateful!'*
>
> (27:38–40)

In this mysterious story it is clear that al-Khidr brought the throne of Bilqîs from Yemen to al-Quds in no time. Because of its great importance in his view of time and the cosmos, Ibn 'Arabî devotes much of chapter 16 in his famous book *the Bezels of Wisdom* to talk about this distinctive phenomenon of timeless motion often associated with the extraordinary actions and manifestations of the saints (*awliyâ'*) (Corbin 1969: 224–226). Ibn 'Arabî, however, declares that in reality this is not motion, but, rather, the throne was annihilated or extincted from its original place and instantly re-created in front of Solomon, in no time, based on the affirmation in those verses that Solomon saw the throne placed 'steadily' before him *before 'he returns his gaze'* (i.e. her throne was not shaking, as one would expect when an object moves quickly and then stops suddenly, owing to the deceleration). So indeed there was no motion, but rather – as al-Qâshânî and Bâlî Effendî mentioned in their commentaries on this chapter – all al-Khidr did was to switch the throne from its original location to the new place in front of Solomon.[6]

From this perspective this kind of phenomenon is easily explained, on the basis of the re-creation principle, since, as Ibn 'Arabî frequently affirms, such apparent 'miracles' do not break the divine laws underlying the phenomena of nature, but only break our 'habits' of perception and expectation [II.374.27]. In fact everything that happens in Nature must be explainable under some divine laws, and is in principle possible to be repeated [II.374.11].

In terms of modern physics, this story actually corresponds to the pervasive phenomena that happen in the atom whenever an electron moves from one orbit to another, when it emits or absorbs a photon of light. It is well known in physics that when the electron absorbs or emits a photon its distance from the nucleus in the centre of the atom changes. This change of the distance, however, is abrupt: i.e. it does not travel through between the two places, but rather it disappears from its original orbit and reappears in the new orbit. Usually such kind of instantaneous motion is scarcely explained according to the novel equations of Quantum Mechanics, and it is not explainable at all with the classical Newton's equations (see section 1.3). Indeed such inexplicable electron behaviour was the driving force behind the discovery of the successful theory of Quantum Mechanics.

If we recall Ibn 'Arabî's definition of motion discussed in section 2.6, we can easily account for this type of timeless motion. Such motion, according to Ibn 'Arabî, is not an infinitely gradual change of positions from the start to the destination, but rather a new creation in the second place. However, if we want to be very accurate, the motion of the throne of Bilqîs, like the motion of electrons between orbits and indeed any elementary motion, takes a single 'Day of event' which is equivalent to a single 'atom of time' of what we may count. Measuring the time that the electron needs to switch between two successive orbs might be the key to testing Ibn 'Arabî's cosmological model and developing it in terms of modern scientific theory. This smallest instant of time is the quantum (*al-zaman al-fard*) of the quantized time that we have discussed in sections 2.8, 4.1 and 6.5.

In normal cases, of course the distance travelled in this smallest moment is also very short, which causes us to see gradual changes; but in principle the distance involved could be anything; as Allah wills, since, whether the change in positions is small or large, the same process of re-creation is taking place. Therefore, what we call the laws of physics or mechanics, are – in terms of Ibn 'Arabî's cosmological conceptions – just a qualitative 'description' of the normally observed order of re-creation, not a real description of the actual process of perpetual re-creation. We shall see further below that this conception has a great impact on our understanding of causality the most fundamental principle in physics.

7.3 Simultaneity

We have seen in section 1.3 that many classical theories considered time as an absolute quantity. This coped very well with the common concept of simultaneity, where events which occur simultaneously in one frame of reference were considered to have occurred simultaneously also in all other frames. With the advent of the Special Theory of Relativity, the idea that light travels at a finite speed in all directions and in all frames of reference changed this piece of common sense. According to this new theory, simultaneous events in one frame of reference are not necessarily considered simultaneous with regard to another frame of reference moving at a relatively high speed with regard to the first.

According to Ibn 'Arabî's view of time and his model of the cosmos that we have described above, the concept of simultaneity will have an even more relative aspect. With regard to us – i.e. considered as partial monads present on the level of multiplicity – it is possible to have simultaneous events. The reason is simply because normally we exist only for at one single location of the whole momentary 'Day of event' (as we explained in section 2.16). For us, at every single moment of re-creation there is a still picture (which contains infinitely many events) displayed in the world. So from this perspective the concept of simultaneity appears like the classical definition.

But according to the re-creation principle and the oneness of being discussed in Chapters 5 and 6 – in addition to the concept of the single Day of event discussed in section 2.8 – there can be no two cosmic 'events' (englobing all of creation) actually happening at the same time, because '*each Day He is upon some (one, single) task*' (55:29). Therefore, in reality there is no such thing as 'simultaneity' – with regard to the Single Monad who is creating the real flow of time (see also Chapter 4) – because It wears only one created form at each instant of time. Simultaneity, and therefore multiplicity, thus appear to occur only because of the re-creation. But in reality there are not any two separate (all-encompassing) 'events' happening at the same created instant of time. We shall see the importance of this conclusion more clearly when we discuss the EPR paradox below.

7.4 Zeno's paradoxes

Motion, as a manifestation of causality, is the main concern behind all the theories of physics, from the pre-Socratics through Newton's theory of gravity to the most recent theories of Quantum Mechanics and Quantum Gravity. Yet there are a number of famous philosophers who have doubted that there could be any motion at all, despite our daily experience. As is well known, those philosophers expressed perspectives similar to Ibn 'Arabî's. Most notably, Parmenides of Elea (b. 510 BC) affirmed cosmological conceptions remarkably similar to Ibn 'Arabî's doctrine of the oneness of being: he held 'the One' unchanging existence to be alone true, while multitude and change were said to be an appearance without reality. This doctrine was defended by his pupil Zeno (b. *c.* 488 BC), whose philosophy of monism claimed that the many things which appear to exist are merely a single eternal reality which he called Being (a term Ibn 'Arabî also applies to the Single Monad). The complex and rigorous adaptation of Parmenides' hypotheses in Plato's *Parmenides* – constantly elaborated by the later Neoplatonists – offers even closer analogies to Ibn 'Arabî's overall ontological system. Zeno wrote a book containing forty paradoxes, and, although his book was lost, four of those paradoxes were discussed by Aristotle in his *Physics*: the Dichotomy, the Achilles, the Arrow and the Stadium. Each of those four paradoxes challenges all claims that there is real motion (Heath 1981: 273–283; Sorensen 2003: 44–57; Darling 2004: 351; Leiber 1993: 77; Erickson 1998: 218–220).

The Dichotomy paradox concludes that there is no motion because that which is moved must arrive at the middle of its course before it arrives at the end. In order to traverse a line segment it is necessary to reach its midpoint. To do this, one must reach the one-fourth point; to do this, one must reach the one-eighth point, and so on *ad infinitum*. Hence motion can never begin, because the sum 1/2 + 1/4 + 1/8 + . . . equals one, but only after an infinite number of additions, and therefore it actually approaches one but never reaches it. Even more perplexing to the human mind is the attempt to sum 1/2 + 1/4 + 1/8 + . . . backwards: for we can never get started, since we are trying to build up this infinite sum from the wrong end!

The paradox of Achilles attempts to show that, even though Achilles runs faster than the tortoise, he will never catch her! Let us suppose that Achilles runs at ten metres per second and the tortoise at only one metre per second, and that when the race started the tortoise was ten metres ahead. After one second Achilles would arrive at the point where the tortoise was when the race started, but the tortoise would have moved one metre further – so that by the time Achilles covers this one metre, the tortoise would have advanced again 0.1 metre, and so on. Thus Achilles can never catch the tortoise.

Zeno bases the above two arguments on the fact that once a thing is divisible, then it is infinitely divisible. One could counter the above two paradoxes by postulating an atomic theory in which matter (or space) is composed of many small indivisible elements. However the remaining two paradoxes cause problems only if we consider that space is made up of indivisible elements that may be cut in indivisible durations of time.

Turning to the third paradox of the Arrow: if we consider the path of an arrow in flight, at each instant of its path the arrow occupies some position in space; this is what it means to say that space is discrete. But to occupy some position in space is to be at rest in this position. So throughout the entire path of the arrow through space, it is in fact at rest! Or if in an indivisible instant of time the arrow moved, then indeed this instant of time would be divisible (for example, in a smaller instant of time the arrow would have moved half that distance).

The fourth paradox of the Stadium is a little more complicated, but it leads to the same result as the above – i.e. that time and space can not be discrete – while, on the contrary, we have seen that the first two paradoxes may be resolved only if we assume that time and space are not continuous

The above four paradoxes challenge not only all scientific theories of motion but also our everyday experience. For this reason they have often been dismissed as logical nonsense. Many attempts, however, have also been made to dispose of them by means of mathematical theorems, such as the theory of convergent series or the theory of sets. Aristotle did not fully appreciate the significance of Zeno's arguments, since he called them 'fallacies', without actually being able to refute them. Many modern scientists like to believe that axiomatic mathematics has dispelled Zeno's paradoxes, where now it is possible to talk about limits and infinity without reaching any mathematical contradiction and it can be proved that the sum of an infinite number of halving intervals is finite. But some recent philosophers such as Bertrand Russell persisted with such arguments, and recently similar puzzling phenomena (called the 'quantum Zeno effect') have been observed in radioactive atoms (Misra and Sudarshan 1977: 756; Grossing and Zeilinger 1991: 321–326).

With Ibn 'Arabî's re-creation principle, we would have no difficulty at all in resolving Zeno's paradoxes and reconciling his conclusion that there is no motion with our daily perceptions. So although there is no real motion in the sense that the object gradually leaves its position to a new one, but rather it is re-created in ever new positions so that we imagine it moving between these places. For example, when we watch a 'movie' on the television, we have no doubt that nothing really moves on the screen, but it is only a succession of different frames. According to Ibn 'Arabî's cosmological perspective, the whole world is exactly like that (see section 6.8). As we noted in section 2.6, Ibn 'Arabî plainly stated that the object that we see moving actually is re-created in the distinct places between its start and destination one after another and does not really 'move' between them [II.457.31], so there is never any real motion in such a way that the object 'gradually moves' along its path. Thus Ibn 'Arabî, like Zeno and Parmenides, believes that the whole world is a manifestation of a single entity which alone can be described to have a real existence. But Ibn 'Arabî's distinctive contribution is to show how the multiplicity of the world emerges from or within the Single Monad.

7.5 Discreteness and continuousness

There is no doubt that Zeno has presented a deep problem which, despite centuries of efforts to resolve it, still seems to lack a truly satisfactory solution. As Frankel wrote:

> The human mind, when trying to give itself an accurate account of motion, finds itself confronted with two aspects of the phenomenon. Both are inevitable but at the same time they are mutually exclusive. Either we look at the continuous flow of motion; then it will be impossible for us to think of the object in any particular position. Or we think of the object as occupying any of the positions through which its course is leading it; and while fixing our

> thought on that particular position we can not help fixing the object itself and putting it at rest for one short instant.
>
> (Frankel 1942: 1–25, 193–206)

This basic dilemma of discreteness and continuousness has kept coming up in various guises, but most clearly in the long historical debate on the nature of light: whether it is particles or waves. With the success of the wave theory in the nineteenth century, the continuum seemed to have won. But in 1899, when Max Planck solved the 'black body problem'[7] by postulating that atoms could absorb or emit energy only in discrete amounts, the age of Quantum Theory began. Soon after that, Bohr used the concept of quantization to construct the first successful atomic model, and Einstein was able to analyse the photoelectric effect only by adopting the quantum nature of light. However, Quantum Theory was not able to solve the question of motion and change, where the continuous theory of Relativity was more successful.

So the human mind is accustomed to classifying quantities as either countable or uncountable, or either discrete or continuous; there is no other way. This is inevitable on the level of multiplicity. But on the level of oneness (i.e. of all-inclusive *ahadiyya* or 'unicity') there would be no meaning for such terms. A first look at Ibn 'Arabî's model could conclude that, on the level of multiplicity, the world should be certainly discrete, and therefore that Ibn 'Arabî might easily adopt the atomist view. But the issue this raises is quite similar to what we have discussed earlier in Chapter 2 about the length of the moment and whether it is composed of discrete sub-moments, or whether it has a length at all. We have seen that it is not easy to decide for either case. Similarly, it is not easy to judge – even on the multiplicity level – whether the world is ultimately continuous or discrete. Although there are discrete events happening in discrete times, still the change from one event to another looks continuous, just like the flow of normal days; there is no abrupt change. Although we can easily divide events over days and classify them according to the date, actually the relation between any two consecutive events that happened during the day is not different from those which happened also consecutively but on different days – for example, right before and after morning or evening. In other words, the motion of the Earth around its axis, though generating the appearance of different distinct days, it is itself a continuous process. Likewise, the all-creative 'motion' of the Single Monad is also a continuous process in everlasting alteration between 'daytimes' and 'night-times', manifestation and being hidden, material and spiritual – yet there is no point of separation or abrupt transformation between any two periods or states. That is why Ibn 'Arabî calls the terms of discreteness and continuousness 'disconnected' (*munfasil*) and 'connected' (*muttasil*), because for him the actual process of change (re-creation) is like a one-dimensional flow of divine manifestation. So if there is an apparent continuity or discontinuity, that would only be in our imagination or abstract consideration, but not in reality [III.324.35–325.18].

7.6 The EPR paradox

Physics theories, which are mainly based on explaining motion and its relation to time, have worked fine for centuries, so that people have been able to send spaceships to the Moon and other planets with great timing accuracy. Dismissing Zeno's paradoxes, therefore, had no negative consequences for applied physics. However, no single theory up to now has been able to explain all observations, especially when it comes to the microcosms of subatomic particles, where new paradoxes are still arising.

One of the most pertinent tests of Ibn 'Arabî's cosmological model is that through its principles the prominent Einstein–Podolsky–Rosen (EPR) paradox could be so easily and readily understood. The EPR paradox demonstrates the discrepancy between the two principal theories of physics: Quantum Theory and Relativity. Quantum Theory is one of the most successful theories of science. It explained the structure of atoms, the properties of materials, elementary particles and stars. Although it was generally consistent with the results of many decades of experimenting, the basic conceptual foundations of Quantum Mechanics can lead to some puzzling paradoxes and strange unacceptable features. The EPR paradox is possibly one of the most compelling of these apparently peculiarly unacceptable features.

In 1935 Einstein, Podolsky and Rosen – in their famous article titled: 'Can Quantum-Mechanical Description of Physical Reality be Considered Complete?' (Einstein *et al.* 1935: 777) – developed a thought experiment to demonstrate what they felt was a lack of completeness in Quantum Mechanics. This so-called 'EPR paradox' has led to much subsequent and still ongoing research. The purpose of the EPR thought experiment was to expose the profound peculiarities of the quantum description of a physical system extended over a large region of space. It seemed that, under certain conditions, a quantum system of two entangled particles could in theory exchange information instantaneously or, at least, faster than the speed of light. This clearly contradicts the principle of 'locality' in Einstein's theory of Relativity, which supposes that the speed of light is a maximum terminal velocity. The phenomena of entanglement also lead to the violation of Heisenberg's sacred 'uncertainty principle', which declares that not all the classical physical observables (e.g. position and momentum) of a system can be simultaneously known with unlimited precision, even in principle.

On the basis of these contradictions, EPR refused this deterministic nature of Quantum Mechanics and postulated that the existence of 'hidden variables', some thus far unknown properties of the system, should account for the paradoxical discrepancy. Niels Bohr, on the other hand, favoured the view put together in the 'Copenhagen interpretation' of Quantum Theory and refused the idea of hidden variables (Bohr 1935: 696).

In 1964 John Bell proposed a mechanism to test for the existence of these hidden variables, and he developed his 'inequality principle' as the basis for such a test (Bell 1966: 447). This was followed by many experiments to verify this principle. The most successful of these experiments was performed by Alain

Aspect and his colleagues in 1982 (Aspect 1982a: 1804; Aspect 1982b: 91). Their experiment consisted of light polarization measurements made on pairs of photons, moving in opposite directions, emitted simultaneously in single transitions by calcium atoms, then measured by sensitive detectors on each side. The results of the experiment clearly violate Bell's inequalities, eliminating the need for the existence of hidden variables and thus supporting the predictions of non-local Quantum Mechanics (which violate Einstein's locality principle and even common sense). The results showed that apparently the left-hand detecting apparatus was sending some kind of message to the right-hand photon informing it as to how the left-hand one was set up, so that the right-hand photon could interact in the appropriate way with the right-hand apparatus. But as a matter of fact there is no communication at all between the left and the right photons, and, if we suppose there is, then the signal would have to travel faster than the speed of light.

Many even more accurate experiments have been performed after that, and all show that it is as if time 'stops' between the pair of entangled particles and they interact instantaneously despite the large distance between them. Although this experimental outcome supports the fundamental concepts of the theory of Quantum Mechanics and contradicts Einstein's locality principle, but there is no adequate theoretical explanation for it, so far.

Now one of the striking consequences of Ibn 'Arabî's principle of re-creation by the Single Monad is that such strange instantaneous behaviour between the two entangled particles would be natural and very easily explained. According to his cosmological theories the constant creation of the manifest world proceeds like a movie composed of succeeding still pictures. Each momentary 'picture' of the world is like a closed system, where any change in one part of this picture would require another synchronizing change or changes in other part(s), so that the whole does not change because there is no external interaction, only 'internal' changes. We have also explained that all 'parts' of the world are created in series by the Single Monad that constantly creates ('puts on') ever new forms, one total individual form at each instant of time – such that each created form then ceases to exist (intrinsically, not through any other force) the next moment after its existence.

Now if we take that theory of cosmic re-creation into account, we can simply say that the two entangled particles in these experiments, like any two entities in the world, never existed together at the same. Rather, the Single Monad first wears the form of the first particle – i.e. causes this particle to exist (under special conditions); then this unique Single Monad itself wears other forms in a specific sequence (see section 5.6), until it comes to wear or create the form of the second particle – regardless of where it is in space. But by the moment of this second state of creation, the first particle is out of existence, and therefore it encountered no time. Now, because the two particles are 'entangled' (in a closed system), any change on the first form that corresponds to the first particle will be kept in the 'memory' of the Single Monad, so that when it comes to wear the form of the second particle it does so in a way that keeps the total state of the system of the two particles unchanged (according to the Quantum Mechanical laws, because it is a closed system). This process is instantaneous; no matter how far apart the two particles are, because only one particle really exists at a time, and during the interval between creating the two particles they were both out of existence, encountering no time. This hypothesis corresponds to Ibn 'Arabî's account of 'the real flow of time' that we explained in Chapter 4.

In fact what we have just said applies to any (large or small) system and not simply the system of those two entangled particles, but in normal cases the effect of the ongoing process of cosmic re-creation is not noticeable because of the many possible changes that could happen in any part of the complex system and the corresponding distraction of our limited means of attention and perception. In other words: any change in any part of the world will cause synchronization changes in all other parts, because the world as a whole is a closed system. This last statement is in fact another and more precise depiction of the whole cosmic process of existentiating causality. This conception also provides a hypothetical explanation for certain 'para-psychological' phenomena.

7.7 Causality and induction

As Ibn 'Arabî often pointed out in his analyses of our unconscious reliance on our 'habitual' forms of perception (*'âda*), the main obstacle that prevents us from discovering the reality of creation is our deep trust in causality and induction. It is true that we live by causes and results, but indeed this is only a limited and superficial perspective with regard to the underlying realities. The fact of the matter is that what we refer to as 'cause' and 'result' – i.e. what Ibn 'Arabî normally calls the *asbâb* (s. *sabab*), or intermediate or 'apparent' causes – are chronologically arranged, but ontologically unrelated. This is quite evident in terms of the re-creation principle discussed above, but Ibn 'Arabî also clearly declares that 'Allah creates things "next to" (*'inda*) the causes and not "through" (*bi*) them' [II.204.13, *Tadbîrât*: 312]. As we've already seen above and in section 2.6 that motion is only a new creation in a different place, so there is no deeper, inherent ontological relation between the two created states before and after motion.

Also – as a result of the readily observed systematic causes and effects – we normally rely upon induction in drawing most of our conclusions about causality. But in chapter 56 of the *Futûhât*, Ibn 'Arabî argues that 'in reality induction (*al-istiqrâ'*) does not give any (true divine) knowledge' [I.285.3]. The reason why he concludes that is simply because nothing is ever truly repeated [I.285.28], as we have also shown above in discussing the re-creation principle. Induction ultimately means that we expect something to occur in a specific definitive way, based on previous observations of similar circumstances and regularities.

For these reasons, according to Ibn 'Arabî, the basic observational and logical principles that the scientists and philosophers of his time relied upon in seeking the truth are indeed not firm, so that their practical success deceives us with regard to their limitations and restricted spheres of application. For this reason Ibn 'Arabî's answer to Ibn Rushd (see section 5.1) was first 'yes', but then he immediately added again 'no', to highlight the limitations of the philosophers' methods, in comparison to his own reliance on divine inspiration and 'unveiling'.

Ibn 'Arabî, however, does not deny the phenomenological relevance of the apparent, intermediate 'causes' (*asbâb*): on the contrary he affirms that they are intentionally established by Allah, so they are unavoidable [II.653.11], and we must rely on such causes to reach our goals. So those observable secondary or apparent causes may not be removed:

> And we already informed you that apparent-causes (*asbâb*) are divine veils, so that removing them is not possible except through them (i.e. by other secondary causes). So the very removing of causes is in fact fixing them; and the reality of abolishing them is (actually) affirming them.
>
> [II.553.31; see also I.382.26, III.340.9, IV.275.23 and III.235.28]

So since they are veils, we have to look *through* them to see the Real, Who continually creates the causes and the results and arranges them in this specific 'habitual' dependency. As Ibn 'Arabî explains, Allah creates simply by the Command 'Be', and therefore He does not need the secondary causes; but He establishes these apparent causes for us to unveil them [II.413.35], though we would never be able to unveil the causes completely and permanently, especially since our own unenlightened mentality itself is the main veil [II.553.6].

In the short chapter 252 of the *Futûhât* Ibn 'Arabî analyses the known spiritual state of 'abolishment' (*al-mahw*), in which the spiritually realized person is freed from the unconscious limitations of habitual perception (*'âda*), so that the rule of apparent 'causality' is removed for him and he can directly perceive the universal reality of ever-renewed creation. There he says:

> For the person in the state of 'abolishment' (of the 'habitual' perception of causality), relying on causes is removed, but *not* the causes. For Allah never deactivates the rule of (divine) wisdom with regard to the (created) things, and the apparent-causes are (like all other creations) divine 'veils' established.
>
> The greatest of those veils may not be removed, which is your own individual self (*'ayn*: also, 'eye'). For your individual self is also the cause (still *sabab*) behind the existence of (the possibility of our) knowing and recognizing Allah – since knowing Allah may not come to exist except through your individual self!
>
> So it is not possible for you (i.e. the veil of your ego-self) to be 'removed' with Allah's wanting to be known (by you). So He abolishes you from (witnessing) yourself, so you no longer stop with (witnessing) yourself, even though your individual self still exists, since the manifestation (of His transforming) influence is from Him. This is just as (in the famous incident at Badr alluded to in verse 8:17) He 'abolished' (the normal causal role of) the Messenger of Allah, may Allah have peace and mercy upon him, with

> regard to his throwing, despite the (actual) existence of throwing by Him: so He said '*and you threw not*', so he abolished him; '*when you* threw', so He affirmed the apparent-cause; '*but Allah threw*' (8:17), and He only threw by the hand of the Messenger of Allah, may Allah have peace and mercy upon him.
>
> [II.553.5]

The apparent-causes are indeed manifestations of the influences of His Names, which is why we see the effects following from those causes, but indeed the effect – like all of creation – is also only (ultimately) caused by Allah Himself. This means, as Ibn 'Arabî affirms, that the names of causes are in fact Names of God:

> Then you have to know that the contingent things (creatures) are by essence in need (of being given existence), so that need always accompanies them because their essence (as contingent, 'needy' things) is forever. So He established for them the apparent-causes (*asbâb*) through the presence of which they receive what they need. So they are in need of the apparent-causes, therefore Allah made the individual-reality (*'ayn*) of those apparent-causes Names for Him. So the names of the causes are from (the influence of) His Names, may He be Exalted, so that (the contingent thing) should only need Him, because this is the correct knowledge (of the nature of things).
>
> So there is no difference, for the people of unveiling, between the names that are said to be the Names of Allah according to (religious) custom and the revelation, and the names of the apparent-causes, so far as their (all) being Names of Allah, because He said: '*you all are in need of Allah*' (35:15). For we see in reality the need for the apparent-causes, so the names of the apparent-causes must be Names of Allah, the Exalted. So we call out in prayer to those names (of the apparent-causes) by (the needfulness of our actual) state, not through outward words (of prayer).
>
> Thus when we are touched by hunger we rush to seek the food that may remove the pain of hunger. So we are in need of it, and it has no need of us – although we (in fact) are only in need of Allah. So that (food) is one of His Names: I mean the form of this food descends to take the place of the spoken or written word of the (corresponding) divine Name. That is why He ordered to be thankful to the apparent-causes (31:14), because He ordered us to be thankful to Him – that is, to thank Him through them.
>
> [III.208.7]

But seeing through the apparent causes (to the One real Cause or Creator) is not easy, and ultimately the meditations and path of ascension of the spiritual seeker are all directed toward that. At the end of this path, the seeker will see only the Image of the Real (rather than his ego-self), and that is why 'whoever knows himself knows his Lord'.

However it is very important to notice that at the highest steps of this ascension, as the seeker advances, he gradually loses the awareness of his limited physical composition, so that what remains at the end is the pure receptivity of 'hearing', just as 'hearing' (*samâ'*, or *sam'*) was the first condition of all things created. Thus Ibn 'Arabî says:

> He said: '*and Our word unto a thing, when We intend it, is only that We say unto it "Be!", and it is*' (16:40), but He deafened us from perceiving this word except by the way of faith, and He blinded us from seeing His intention (*tawajjuh*) on creating the things, by the apparent-causes that He established. thus He sends the rain down, and it falls; and He makes fertile the land, the seeds sprout; then He sends out the sunlight, so the plants rise, and (the seeds) harvested, ground up, kneaded, baked and chewed by the teeth; then they were swallowed, ripened in the stomach, taken by the liver and made into blood, sent into the vessels and distributed over the body until the (life-giving) vapour rises out of them, so the life of this body is for the sake of that breath (that we breathe).
>
> So these are the primary apparent-causes (for life), along with (His) moving the celestial spheres and the motions of their planets, casting the rays on the places (illuminated by) the (celestial) lights – (all that) under the supervision of the Universal Soul, with Allah's permission, and the support of the (First) Intellect for (the Universal Soul). All these are veils, established as the fundamental (apparent-causes), in addition to the other, lesser apparent-causes. So the (human receptivity or) 'hearing' needs to break through all those veils until it hears the (creative divine) word 'Be!'. Therefore He created in the person of faith the power of faith, so that it flows into his hearing until he realizes the word 'Be', and it flows into his sight until he witnesses the One Who gives existence to the apparent-causes. And He has done all that through the 'Breath of the All-Merciful'.
>
> [II.413.35]

So – despite their ultimately being ontologically unrelated – one cannot deny the effect of the apparent-causes. Therefore, we have to look at other prospects for redefining the principle of causality. We have seen before (section 3.5) that Ibn 'Arabî affirmed that after all the forms in the world have been created in the six Days from Sunday to Friday (which is space), nothing remained to be created on Saturday, that is the 'Day of eternity' or the instant of time that we live in at each now. What remains is only continual creative changing the states of things 'from (one) state to (another) state and from (one) station to (another) station' [I.61.14]. Therefore the system of the world, as a still picture created in the Week, is a 'closed system' in the meaning that all changes in it are necessarily internal changes.

Therefore, whatever change happens in one part of this world will require another change in other part(s), so that the total state does not change. This last statement seems to articulate a new, more adequate form of the causality principle, and indeed this is exactly one of the principles of Quantum Mechanics which has proved to be successful over the last century (see section 1.3). This is in fact a very important conclusion and it could be the key to a new understanding of the reality of motion. This also might explain the types of what science tends to view as inexplicable 'para psychological' phenomena, such as the telekinetic effects and telepathy which are widely known as *karâmât* (acts of grace). The world in all its directions of space and time, past and future, here and there, is already encoded (via the divine Foreknowledge) in the Single Monad, so whoever has any access to that Source should be able to predict what is in or affects other parts of the world. If we want to study this issue further we have to look at the divine Attributes of Power and Will, and their image in the human being and the world, and relate them to time and space, as Ibn 'Arabî discussed at the end of his short treatise *Al-Durrat Al-Baydâ*.

7.8 Superstrings and the science of letters

In the standard model of elementary particle physics, particles are considered to be points (or spheres) moving through the four dimensions of space-time. Extra abstract dimensions are needed to take into account the different properties such as mass, charge and spin.[8] This standard model eventually led to obvious discrepancies between Einstein's theory of General Relativity and the Quantum Field Theory that is essentially based on the wave properties of matter (see section 1.3, and also section 7.6).

Around 1985, the new String Theory suggested that all elementary particles can be represented by fundamental building blocks called 'strings' that can be closed, like loops, or open, like a hair. The different vibrational modes (or 'notes') of the string represent the different particle types, since different modes are seen as different masses or spins. One mode of vibration makes the string appear as an electron, another as a photon. But one of the most remarkable predictions of String Theory is that space-time has ten dimensions rather than four. However, six of these dimensions are curled up very tightly, which is why we may never be normally aware of their existence. Other subsequent extensions of the String Theory anticipate even higher dimensions.

There are deep and exciting similarities between the principles of the String Theory and Ibn 'Arabî's views. We have already mentioned in section 2.1 that he says that there are four 'fundamental principles of existence' that are – in addition to 'another six derived from them' – enough to describe the state of everything in the world [III.404.22]. But what is most exciting in this regard is Ibn 'Arabî's concept of the mysterious 'science of letters' (*'ilm al-hurûf*) or what he calls the 'world of Breaths' (*'âlam al-anfâs*).

We've already mentioned above (section 6.6) the complex symbolic cosmological analogies that Ibn 'Arabî elaborates, beginning in the long second chapter of the *Futûhât*, between the cosmos, the Single Monad and the Greatest Element on the one hand, and the world of letters on the other hand. Ibn 'Arabî adds that 'the world of letters is a nation like other nations . . . and those who know that are only the people of unveiling in our path, . . . and they (the nation of letters) are grouped into groups like the normal world that we know' [I.58.12].

These 'groups' refer to the groups of the spiritual hierarchy that we have mentioned above, which Ibn 'Arabî explained in detail in chapter 73 of the *Futûhât*.

Ibn 'Arabî also mentioned at the end of chapter 1 of the *Futûhât* that 'the first line that I read (from the hidden knowledge of the Spirit from whom he took everything that he wrote in the *Futûhât*), and the first secret of this line that I knew, is what I am going to mention in this second chapter' [I.51.30]. Then he wrote about 40 extremely dense pages on the cosmological dimensions and significance of the science of letters in chapter 2.[9] In this chapter and other parts of the *Futûhât*, Ibn 'Arabî mentioned many mysterious facts about the letters and their world and cosmological meaning. For example, Ibn 'Arabî explained there the relation between the characters of the word *azal*, as written in Arabic, and the meaning of time (*zamân*) by tracing the mysterious relation between the letters in both words, as he mentioned in his *Kitâb Al-Azal*.

This is in fact a very broad and complicated subject, and we cannot go into details here. What we do want to summarize is that Ibn 'Arabî considers the entire cosmos as the words of the Real spoken through the 'Breath of the All-Merciful' [I.366.1, II.403.21, 459.6; see also *Al-Masâ'il* 105], just like the meanings that we create through the words that we speak which are also composed of letters (or sounds) that are essentially the vibrations of our vocal strings under the influence of our breath. In chapter 2 of the *Futûhât*, Ibn 'Arabî gives details about the cosmological significance of each letter, how it is produced, what kind of vibrations it carries, and also the different orbs that contribute to produce it. Then in the long chapter 198 [II.390–478], which is titled 'On Knowing the Breath', Ibn 'Arabî mentioned remarkable facts about these cosmic meanings of the letters and sounds, and he explained the role of each divine Name of Allah in creating the different parts of the world and the different letters of the alphabet.

As one small illustration, we refer here to the letter (and sound) *alif* (ا), the first letter in the Arabic alphabet (and many other languages), which Ibn 'Arabî treats as symbolically identical to the Single Monad we have mentioned above (section 6.6) – not only because it is first but because it represents the closest thing to the pure creative, foundational divine 'Breath' itself. First Ibn 'Arabî asserts that '*alif* is not from (other) letters' [I.65.23], but he stresses that 'all letters (like the world) may be broken down into and built up from it, while it does not break down into them' [I.78.22], so this letter *alif* is present in every letter or word, just like the Single Monad that is also present in everything in the world. Indeed any sound that we produce starts by the sound of letter *alif* because it is simply the beginning of the blowing of the breath through the larynx.

So since the cosmos is the words of the Real and those words are composed of letters or sounds produced through the Breath of the All-Merciful;, these letters are the strings that constitute everything in the cosmos, just as the meanings that we create when we speak are also composed of the letters of the alphabet. Even the written shape and curvature of the Arabic characters, for Ibn 'Arabî, have deep hidden meanings that relate to the cosmos in many mysterious ways: in that sense, those shapes, just like the strings in the Strings Theory, are essentially either open like letter *alif* (ا), or closed like letters *mîm* (م) and *wâw* (و).

The science of letters and of their equivalent numbers (*'ilm al-jafr*) was not Ibn 'Arabî's own invention, but was widely known in various 'esoteric sciences', for example those that deal with magic and talismans, where they replace each letter by its equivalent numbers and make certain calculations and tables that are said to have secret magical effects, or may tell hidden facts. In her famous book *Mystical Dimensions of Islâm*, Annemarie Schimmel devoted a separate appendix to the wider theme of letter symbolism in Sufi literature (Schimmel 1975: 411–425). In fact this kind of mythology dates back to the time of Pythagoras of Samos (582–504 BC), who visualized the world as perfect harmony, like musical notes, which depends on the system of numbers (which were written with the same letters of the alphabet in both Greek and later in Arabic).

As we said, these letters are arranged in a symbolic cosmological hierarchy parallel to the spiritual hierarchy of the saints that Ibn 'Arabî explained in chapter 73 of the *Futûhât*. In fact it is noteworthy that Ibn 'Arabî calls the members (*awliyâ*) of the spiritual hierarchy 'the world of Breaths' [II.6.21], or 'the Men of the world of Breaths' [II.11.9]. Also that is why sometimes Ibn 'Arabî calls the single Days of each singular instant 'the Days of Breaths' [III.127.34], because in this Day the creative divine Breath is taken which is the string or the vibration that appears in existence. Hence in each single Day the string (or the Single Monad, or the 'real-through-whom-creation-takes-place') is vibrating to produce the letter (or sound) *alif*; and the world therefore is the words that are composed of these *alifs* that are produced in the succeeding Days. The Single Monad is the ultimate elementary String, but there are also other elementary strings: just as letter *alif* forms other letters (both in writing and speaking) the Single Monad also forms other monads that are the entities of everything in the world.

For Ibn 'Arabî, this cosmological analogy applies both to speaking (sounds) and to writing (characters), because the 'Higher Pen' (that is the Single Monad) is creating the cosmos by *literally* writing the words of the Real in the 'Higher Tablet' of the Universal Soul. This process of writing produces the 'Pen-sounds' (*sarîf al-aqlâm*), which are the vibrations that are referred to in the hadith recounting the ascension of the Prophet Muhammad [III.61.9].

> So we say that the First Intellect that is the first-created (or first-originated: *awwal mubda'*), and he is the Higher Pen. There was nothing else originated before (*muhdath*) him, but he was influenced by what Allah newly originated in him by raising up through him the 'Protected Tablet' (of the World-Soul), like the raising up of Eve from Adam in the world of material-bodies, so that this Tablet is going to be the substrate and place for what this divine Higher Pen writes (through the 'words' of creation). Now the delineation of the letters is designed to indicate what the Real made as signs pointing to Him.
>
> So the Protected Tablet was the first existent raised up (from another: *mawjûd inbi'âthî*). And it is reported in the revelation (of the Prophet) that

> 'the first of what Allah created was the Pen; then He created the Tablet and said unto the Pen: "write!", so the Pen said: "what to write?" Then Allah said unto him: "you write and I shall dictate you"' [*Kanz*: 15116]. So the Pen writes in the Tablet what Allah dictates to him, which is His Knowledge regarding His creation that He shall create till the resurrection Day.
>
> [I.139.23]

Ibn 'Arabî also divides the letters of the alphabet between four existents: the real (through whom creation takes place), the angels, the jinn and the Humans [I.53.1], which we may render into vibrations in 0-D, 1-D, 2-D and 3-D as we shall explain in the following section. This, he explains, is because hearing (*sam'*) is based on four realities [II.367.24], and that is why in the science of music and notes there are four main notes: the *Bum* (the thick string), the *Zîr* (the highest string), *Muthannâ* (duo), and *Muthallath* (trio): each moves the soul in a special way, causing the emotions of happiness and sadness [II.367.26].

But as Ibn 'Arabî explains in a chapter (182) entirely devoted to 'hearing' – i.e. metaphysical 'receptivity' in all its forms:

> So 'Hearing', in this sense, is divided into three kinds: divine hearing, spiritual hearing and natural hearing. The divine hearing is that of the (divine) secrets (*asrâr*) and it is hearing from everything, in everything and through everything, because all of the world, for them (the true divine 'knowers'), *is* the words of Allah, and '*His words are never exhausted*' (18:109, 31:27). Therefore they have, corresponding to those (divine creative) words, 'hearings' that never end . . .
>
> And the spiritual hearing is connected to the sounds (*sarîf*) of the divine pens on the Tablet of what exists, (which is) 'Protected' from changes and substitutions because the whole of existence is '*a spread parchment*' (*riqq manshûr*, 52:3) and the world in relation to it is an '*inscribed book*' (*kitâb mastûr*, 52:2): so the Pens speak out, and the ears of the minds hear, and the words are engraved (in manifest existence) so they are witnessed.
>
> [II.367.7–9, 18–19]

So these words *are* the spiritual and material world that we live in, which is therefore the succession of the vibrations (letters) produced by the divine creative Breath through the Single Monad that is the Universal Intellect, and displayed in the Universal Tablet. This subject is indeed very diverse and important, and it is worth a separate study. Ibn 'Arabî himself spent a good part of the *Futûhât* on this subject, as illustrated in the long chapters 2 [I.51–91] and 198 [II.390–478].

7.9 The properties of matter

In apparent disagreement with physics, and also with common sense, Ibn 'Arabî paradoxically declares that most common properties of matter like weight, density, transparency and softness are related to the perceiver and not to the objects themselves [II.458.14]. There are only two exceptions, the colour and the shape, where Ibn ‘Arabî accepts that they can be related to something in the object itself, though they may also be dependent on the perceiver like other properties.

At a first glance this might be difficult to accept, especially since it clearly contradicts our daily experience. However, we have already seen in section 2.6 that Ibn ‘Arabî's unique understanding of motion may be understood only on the basis of the oneness of being and the re-creation principle that we explained in section 5.6. Similarly, if we accept that objects and the whole world are continuously created and re-created by the Single Monad, then we have to revise our view about the structure of matter: for there actually exist (in this view) only the individual substances or monads and their forms, so that other properties are consequences and not intrinsic.

Regarding the structure of the cosmos, Ibn ‘Arabî also mentions that the structure of the higher world (i.e. the planets, spheres and stars) ‘is different from what the cosmologists say, although what they say is based on (observational) proof(s); and it would have been possible for Allah to have it arranged that way (as they say) – but He did not’ [II.670.7].

This statement – which might allude to such prevalent current astronomical theories as the model of ‘epicycles’, or the assertion of a unique distinctive ‘element’ (‘quintessence’) constituting the higher spheres and planets – means that he regards the models of the Cosmos devised by scientists and philosophers as logically possible, but not true, solutions to the results given by astronomical observations.

7.10 Dimensions of the unseen world

One of the most obvious differences between science and theology is that the first speaks only about physical phenomena (including energy), while the latter assumes the existence of spiritual or non-material beings, such as the jinn and angels, and of various spiritual worlds, including those dimensions associated with the Hereafter. Therefore one of the possibilities in order to bridge that gap is to extrapolate modern laws of physics and cosmology to those unseen worlds. Some recent scientific attempts have been made in this regard, and most postulate that the reason why we do not normally see those supra-natural worlds is that they have higher dimensions (e.g. nine or ten dimensions).[10] However, there are indications that Ibn ‘Arabî sometimes suggests that angels and jinn are ‘physical’ or ‘natural’ creatures less advanced than humans. In that case, angels and jinn have lower dimensions – one-dimensional (1-D) and two-dimensional (2-D) respectively – while we humans, are three-dimensional (3-D) creatures, who in the Hereafter (or perhaps even before) may be developed into four dimensions. Here we are speaking about *spatial* dimensions alone.

Since we are now ordinarily 3-D creatures, we can not in general see jinn and angels because our ordinary sensory tools are used to detect only 3-D phenomena. In contrary to that, jinn and angels can see us – not because their senses are capable of perceiving 3-D phenomena but because they rather perceive and interact with our souls and intellects which are of their own nature (of 2-D and 1-D respectively). Likewise jinn and humans can not in general see the angels. However, it is possible for humans to cross over to the jinn and angel worlds by transforming into 2-D and 1-D [I.168.20] and this is what happens to the seeker at the beginning of his or her path and in the spiritual ascension, which is why advanced human beings are able to speak and interact with spirits easily [III.332.11], until they ultimately witnesses the Real Himself. But in order to reach this final stage, the spiritual seeker must be 'annihilated' into 0-D, because the Real can only be witnessed by His secret ('the real through whom creation takes place') which He has implemented in everything [I.168.22, see also III.540.33].

Dimensions play a very important role in modern cosmology and mathematics. There are real dimensions and abstract dimensions. In principle, we can – mathematically – assign a dimension to any parameter of a specific function. For example the weather at any point on the Earth is a function of many parameters such as time, place, the nuclear reactions on the Sun, the amount of cloud in the area, the direction of winds etc. Each one of these parameters can be considered as an abstract dimension for the sake of simplifying the mathematical study of the dependency of weather on these parameters or dimensions. Real dimensions, on the other hand, are only those three dimensions of space (i.e. length, width, depth; or x, y, z), and no more. Although time is considered as a real dimension in Relativity, it is not a spatial dimension and so we shall not consider it as real in this regard (see also section 2.5).

So here we shall speak only about the three real dimensions of space. In fact those three dimensions make six (as Ibn 'Arabî often points out), if we take into account that each dimension has two directions. Those six dimensions or directions are: (up, down), (right, left), (front, back); or $(-x, +x)$, $(-y, +y)$, $(-z, +z)$.

I have to mention, however, that Ibn 'Arabî never arranged the creations in terms of dimensions as we are suggesting in this section; but this possibility is quite evident from his various texts, and many of his ideas can be easily understood on this basis, as we shall see shortly.

With regard to dimensions, existence can be divided into five categories, as we shall explain in the remainder of this chapter:

7.10.1 The real

He is the Being of 'zero dimension' (0-D): i.e. of no dimensions, which means He is independent of space and time. That is why Ibn 'Arabî sometimes symbolically indicates God as a dot or a point [III.275], for example as the centre of the circle of creation as we have seen in the cosmogonic diagram in Figure 7.1; also, the dot that is used in many characters of the Arabic alphabet, especially the letters *bâ'* (ب) and *nûn* (ن). However, 'the Real' here may refer to Allah Himself or also to the Greatest Element who is the Image of Real, or the 'real through

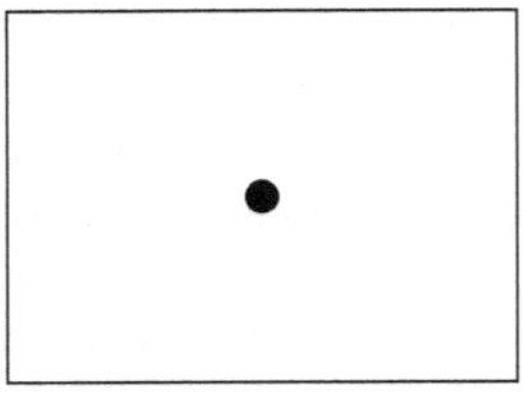

Figure 7.1 0-D, a point. It actually has no length, width or breadth. This is what is called a geometrical point.

whom creation takes place' (*al-haqq al-makhlûq bihi*), and he is the true 'Image of God', as we have discussed in the two preceding chapters. So when we speak about the Real with regard to the creation and spatial dimensions, we usually mean the Greatest Element rather than Allah, since Allah Himself, His Essence, is beyond all descriptions including dimensions.

As we have seen in an earlier diagram in Figure 5.1, Ibn 'Arabî symbolically represents the world as a 'circle' whose centre is always the Real or real. The abstract point in the centre is one unit: i.e. it can not be divided or fragmented, nor can it be even described because it has no dimensions. However, this point faces the infinite multiplicity of the points in the circumference of the circle. Moreover, each point of the circle is similar to or an image of the central point. This is identical to the meaning of the hadith that Allah created the Human Being (and the world, according to Ibn 'Arabî) 'on His own Image' (see section 3.1). Also each point in the circumference can be a centre to a new circle [III.275], just as everyone or everything has his or her own world, in his or her imagination. Also, any point of the circumference can be considered the beginning of it and also the end [I.259.24], so '*He is the First and the Last and Hidden* (centre point) *and Manifest* (circumference)'. And each line that goes out of the central point reaches a point on the circumference. This means that everything originated from the Real (in the centre) and it returns to Him (on the circumference) [II.538.26], which can be understood as the cosmogonic meaning of the verse: '*to Him you return*' (2:28) and *to* '*Him everything returns*' (11:123). Finally, because all the points are in essence a manifestation of the Real, therefore 'the worm and the First Intellect are equal with respect to the essence, but the difference appeared in the form' [III.452.33]. Similarly, as Ibn 'Arabî points out:

> Since the lines (the radii) that go out from the Point in the centre of the circle to the circumference that comes to exist through that Point are *equal* to all parts of the circumference, so likewise the (creative, existentiating) relation of the Real, the Exalted, to the totality of all existents is the same: there is no change in that relation at all. So all things are looking to Him and accepting from Him that (existence) which He bestows on them, just as the parts of the circumference are facing the Point (in the centre).
>
> [I.125.23]

Moreover, any object is composed of a number of points which are similar but have different attributes; the object is the sum of those points. But we can not say that the object is (nothing but) the (particular) point, nor we can say that it is not the point. Similarly, the world is the sum of the manifestations of all the divine Names of Allah, the Real, but we can not say that the world is the Real, nor can we say that the world is not Him [III.275.32].

For Ibn 'Arabî, the Real is creating the world by continuously and endlessly manifesting Himself in different forms and forming the cosmos point by point (in series as we explained above), including the perceived, perception and the perceiver [II.484.22]. Therefore at any single instant of time (the 'now', or the real existence, *hâl*) there exists only Him. This is not like saying that the things *are* Allah, 'because He is He, and the things are the things' [II.484.28], but they are the manifestations of the Names of Allah. In this sense, as we have just seen, 'every name in the world is His Name, not a name of other than Him; for it is the Name of the Manifest in the locus of manifestation' [II.122.14].

The identity or essence of each one of us (i.e. the individual soul) is a point in the circle of Creation, and so is the identity of everything in the world. So we can here clearly see the meaning of the verse: '*We shall show them Our signs on the horizons and within themselves until it becomes clear to them that it/He is the Real*' (41:53),[11] which can be then considered the most obvious Qur'anic basis of Ibn 'Arabî's understanding of the 'oneness of being'. Also, since each point of the circle is similar to the central point, 'whoever knows himself knows his Lord' [I.328.31, II.298.30].

7.10.2 Angels

The angels are beings of one dimension (1-D) and they are made of light (*Inshâ' al-Dawâ'ir*: 27). Angels are the first creatures: the Pen himself is an angel, and he is the first creature and everything else is 'written' by him in the Soul (the Protected Tablet), which is two-dimensional as we shall see further below: so their relationship is just like a normal pen and writing-board, in that they are one- and two-dimensional respectively. Angels occur by the repeated manifestations of the Real (at least two subsequent manifestations); just as the line is composed of at least two points [III.276.3]. Ibn 'Arabî says:

> The reality of the angel does not accept deviation, because he is the origin of the straight line connecting the two nines (of the divine Source and its human receptacles).[12] For bending (from that straight connection) is deviation, and he (the angel) does not have any bending, but he goes back and forth between the straight (creative) motion and the reverse motion (returning from the creature to God). So he is precisely the subtle thread (*raqîqa*: connecting the Source and the creatures) itself.
>
> [I.54.21]

On the other hand, there is strong evidence in Ibn 'Arabî's texts that some angels indeed function as, among other things, forces of Nature: 'They are called angels (*malâ'ika*) because they are links, conductors that link the godly rules and divine effects by material worlds because "*al-malak*" (the angel) in (Arabic) language means the "force" and the "intensity"' (*Inshâ' al-Dawâ'ir*: 27).

And he also says: 'There is no place in heaven or Earth but that there is an angel in it. And the Real continues to create angels from the (creative divine) Breaths of the worlds, as long as they are still breathing' [I.123.2].

It is also known in physics that there are four fundamental forces in nature, which are: the force of gravity, the electromagnetic force, the weak nuclear force, and the strong nuclear force. Nature is built upon these four forces. Those four forces can be conceived as manifestations of the four prime Archangels – *Mâlik*, *Jibrâ'îl* (Gabriel), *Mîkhâ'îl* (Michael) and *Isrâfîl* (Seraphiel) – because Ibn 'Arabî affirms that those four angels are the ones who bear the Throne (*al-'arsh*). Although it is mentioned in Qur'an that '*eight shall, in that Day (of resurrection), bear above them the Throne of thy Lord*' (69:17), he asserts that when this verse was recited before Muhammad, he said 'and they are today (in this world) four' [I.148.2, III.184.28, and also *'Uqlat Al-Mustawfiz*: 43–44], 'and tomorrow (in the Hereafter) they are going to be eight' [I.149.29]. Ibn 'Arabî also explains that *al-'arsh* (usually translated as 'the throne') in the Arabic tongue refers to 'the Kingdom' in addition to 'the Throne' [I.147.33], so, if this verse refers to the divine Kingdom, then its bearers or holders are those who are in charge of its affairs, and these are like the four supports (*awtâd*) who hold up the house. For this reason we find that Ibn 'Arabî also affirms that the four *awtâd* (spiritual Pillars) by whom Allah sustains the Earth get their power from the spirituality of these four angels [I.160.25, II.7.1], though he mentions here *'Azrâ'îl* (the angel of death) instead of *Mâlik* (the master of Gehenna). In the Hereafter, he explains elsewhere, the three prophets Adam, Muhammad and Ibrahim (Abraham) in addition to angel *Ridwân* (the Warehouser of Paradise) shall also contribute, so those eight will be the holders or bearers of the Throne or Kingdom in the Hereafter [I.148.11]. We shall see further below that this helps explain why the Hereafter could be considered as four-dimensional (4-D).

Here Ibn 'Arabî also shows that the Throne *is* the Kingdom, and that it is confined in four things: body, spirit, food and state. So those eight sustainers of the Throne are assigned the various duties as follows: 'Adam and Seraphiel for forms (for bodies), Gabriel and Muhammad for spirits, Michael and Abraham for subsistence, and *Mâlik* and *Ridwân* are for threat and promise (i.e. the states in Gehenna or Paradise)' [I.148.3].

Therefore, if we want to compare those four angels who hold up the divine Kingdom and the four elementary forces which operate in Nature, we can clearly see, for example, a correspondence between gravity and Seraphiel, since both operate upon the forms, the bodies. We can also see clear relations between the electromagnetic force and Michael, because both are responsible for subsistence and food, when we remember that all the food that we eat is in the end produced by light and heat which are electromagnetic waves (forces) emitted by the Sun. It is, however, not easy to establish the relation between Gabriel and Mâlik on the one hand and the weak and strong nuclear forces on the other hand, but we can mention that Gehenna is said in the Qur'an to be '*fuelled by stones*' (2:24, 66:6), and this could mean the nuclear energy that is available inside the atoms for which those two nuclear forces are responsible.

On the other hand there are so many types of angels who also have different states or levels (37:164) and different structures. Some of those various types are mentioned in the Qur'an (37:1–3, 51:1–4, 77:1–5, 79:1–5 etc.), and Ibn 'Arabî talks about them in some detail very often in his writings [III.445.35–446.6, and also *Inshâ' Al-Dawâ'ir*: 27]. And it is quite clear by studying these types that they are assigned specific duties, just like the different forces in Nature (including the elementary and other forces).

Therefore, our suggestion that angels are beings of 1-D appears justified because such natural forces always operate in one dimension (they are also represented as vectors '→' in physics and mathematics), though their effects might appear in two or three dimensions as well.

Notice also that, just as the real (the Greatest Element) is 0-D and can be represented by a dot, the Single Monad is 1-D and can be represented by a line. In letters, the letter *alif* is made by the flow (*sayalân*) of the dot. So also the Single Monad is made by the flow (the repeated creative manifestation) of the Greatest Element (see Figure 7.2).

7.10.3 The jinn

The jinn are beings of two dimensions (2-D) or four directions, and according to traditional sources they are made of fire. Ibn 'Arabî affirms that

> The fire jinn got the four (of letters) because of the facts they are based on that caused them to say as the Real, the Exalted, told: '*then I will approach them from between their hands, from behind them, from their right side and from their left side*' (7:17), and then their facts are over, they have no further fifth fact to seek through it further level, and you should be aware not to think that this is possible for them that to have the height (up) and its counterpart (down) by which the six directions are complete, because the (their) reality does not permit that as we affirmed in the book *Al-Mabâdî wal-Ghâyât*.
>
> [I.53.8]

It is evident, therefore, that jinn can move only in four directions: the plane or the surface. They have no sense of height, so their space is a sub-space of ours,

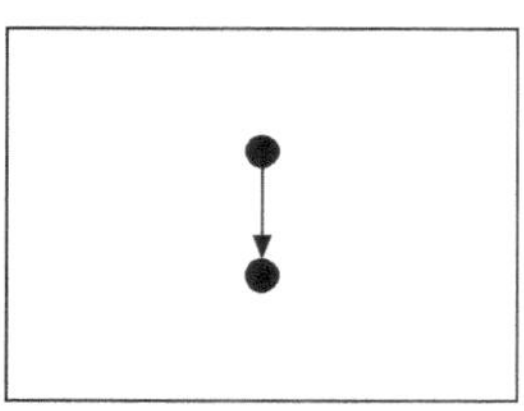

Figure 7.2 1-D, a line segment. It has a length, and it is created by moving a point.

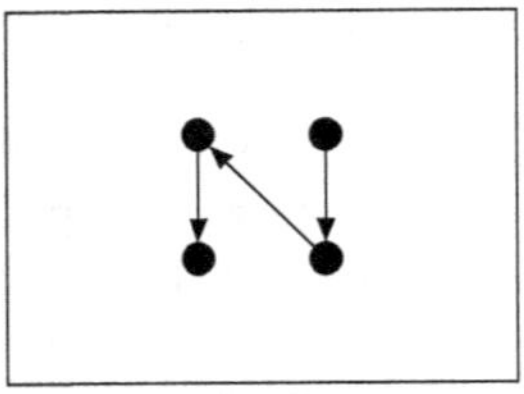

Figure 7.3 2-D, a unit square. It has length and width, and it is created by moving the line segment in a direction perpendicular to the line on which the segment lies.

and that is why we do not see them, they are too thin and tenuous, just like the surface of a heat wave or flame. It is also noteworthy in this regard here that Allah said in the Qur'an that the angels say: '*to Him (to Allah) belongs what is between our hands and what is behind us and what is between that*' (19:64); so He did not mention the sides as in the case of jinn. This therefore supports what we said above about the angels being only 1-D.

In fact, just as the structure of angels is like our spirits, the jinn's structure is similar to the structure of our souls, as Ibn 'Arabî affirms: 'the inner (*bâtin*) of the human being is in fact jinn' [I.85.6], and also it is possible that the inner of jinn is angels and the inner of angels is the real, which is again another form to express the oneness of being.

7.10.4 Humans

Humans of course are beings of three dimensions (3-D) and they are made of earth or clay. As we said above, humans can decompose into 2-D and 1-D so that they may interact with jinn and angels, and they may also decompose into 0-D so that they may witness the Real, but they may not of course become like the Real as God though they may become like the 'real through whom creation takes place' who is the perfect Image of God, or the Perfect Human Being. And as we've said that this is the aim of the Sufi; to decompose into 0-D, which means to purify one's self and get rid of all the earthly (3-D) attachments.

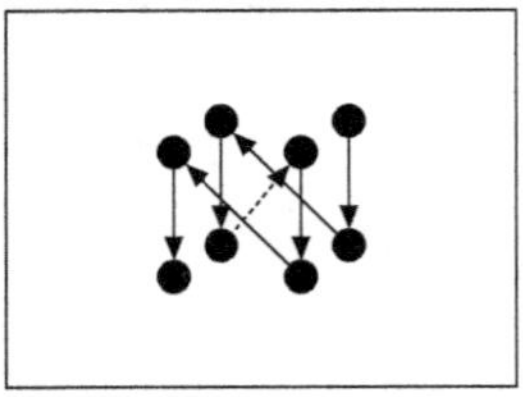

Figure 7.4 3-D, a unit cube. It has length, width and depth. It is created by moving the square in a direction perpendicular to the plane. Notice that seven motions are needed to make up the cube.

7.10.5 The hereafter

We can not talk too much about the Hereafter, because it has been mentioned that it is different from what we may imagine [*Kanz*: 39236, 39241], but as we have mentioned above that the holders of the Throne or Kingdom are now four and in the Hereafter they shall be eight. So it is possible that the Hereafter will be four dimensional (4-D) because doubling the number of forces requires new dimensions for the new forces to operate.

Moreover, we notice that our major senses relate to dimensions in the following manner. Hearing requires only 1-D, because the propagation of sound waves is received by the ear one bit at a time. Seeing, on the other hand, requires 2-D because at any instance we may perceive a picture that occupies a surface. To conceive 3-D, however, we need imagination because the 3-D space that we conceive is built up as a result of the integration of 2-D pictures that we perceive through the flow of time. We perceive 1-D only by hearing, and 2-D by seeing, and we conceive 3-D, by imagination. So our thoughts, or imaginations, that we have in our memory are indeed mostly 2-D pictures (but also 1-D sounds) but by integrating them over time we conceive the volume. Thus, it is possible, and plausible, that in the Hereafter we shall gain new faculty, more advanced than imagination, that allow us to conceive of 4-D. In this case our thoughts will be 3-D as confirmed by Ibn ‘Arabî and many hadith that describe the Paradise. For example Ibn ‘Arabî confirms that people in Paradise shall have the power of creating through the command ‘be’, just as Allah does in this world [*Al-Masâ’il*: 126, I.84.21, II.157.26, II.440.35, II.441.26, III.295.17], and he also affirms that this is also attainable (by some people) in this world [*Al-Masâ’il*: 126, III.295.14]. This is also called *al-fi‘l bil-himma* (doing by intention or determination) [I.259.33].

On the other hand, and since we have seen that the Real is 0-D and that the angels are the first repeated manifestation of the Real and so on, the jinn are manifestations of the second order (in 2-D). Thus humans are more advanced and complex that jinn because they are created by the manifestations of the Real in the third order. Likewise, it is expected that the Hereafter will be the fourth-order manifestation.

The Creation is done through the Pen who is writing the words of the All-Merciful. This Pen started by writing the dot (absolute Spirits, including the real; 0-D) and then the angels by making a line (1-D), and he continued until a certain term where he started another line, thus making a plane (2-D) and that is the creation of jinn. After that and at the certain term also, he started making new planes, thus forming the volume (3-D) that we are in now. And eventually the Pen will start, at a certain term (*ajal*), a new dimension to open the life in the Hereafter (4-D). This ‘term’ is either the time of our death (*al-ajal*) or possibly the time of spiritual realization (*al-fateh*) which is also a kind of voluntary death [IV.354.19] (Chittick 2002: 105–107). We have already seen in section 2.11 that Ibn ‘Arabî specified the starting times between those different terms (see Figure 2.1).

It is also possible in Paradise that the Creation will continue, endlessly, into higher dimensions (*n*-D, where *n* is any integer number), more than four. Maybe that is what it means when Ibn 'Arabî says: 'seeking (*al-sulûk*; into Allah) is always requested in this life and also in the Hereafter. For if there was a destiny, it would have been possible to achieve (but there is no destiny)' [*Al-Masâ'il*: 203]. It is probably here where abstract mathematics – which can deal with any number of dimensions and different types of topological spaces – has anticipated this hypothetical structure of the world, whereas physicists up till now can not conceive of more than three dimensions.

Finally, we must add that conceiving the world in this way according to the dimensions will greatly help in any possible future computer simulations to test the Single Monad model, because in this way we should be able to see whether the current structure of stars and galaxies could have happened starting from the initial conditions and rules that we described above and in Chapter 4 when we discussed the actual flow of time.

Notes

1 Cosmology and time

1 The geocentric view considers the Earth to be in the centre of the universe, while the heliocentric view considers the Sun to be in the centre. Modern cosmology, however, asserts that the universe, being a closed space-time arena, does not have a centre; any point may be considered a centre, just as any point on the surface of the Earth may be considered a centre (with regard to the surface, not to the volume). So whether the Earth or the Sun is in the centre of the universe is a valid question only with regard to the solar system which was the known universe in early cosmology, but it is no longer valid after discovering the galaxies and the huge distances between stars outside the solar system. It is worth mentioning here that Ibn 'Arabî clearly asserted that the universe does not have a centre [II.677.19].

2 For more information about this subject see: Bieńkowski (1972).

3 The redshift is the displacement (towards the red side) of the spectral lines of the light emitted by stars when it is received on the Earth, and this is due to the high speed of the motion of stars away from us. The amount of the shift towards the red is directly proportional to the distance of the star away from us, and this is how distances to far away stars and galaxies are calculated with a high degree of accuracy.

4 For more information about the principles of Quantum Cosmology see Linde (1990), chapter 3.

5 For more information about Ibn 'Arabî's life and intellectual background see Addas (1993) and Hirtenstein (1999).

6 For a full list of books and manuscripts attributed to Ibn 'Arabî' see: Yahya (1964). In this book Othman Yahya mentions over nine hundred books (with about 1,395 titles) attributed to Ibn 'Arabî. Most of them however, as Yahya shows, are not really by him, and also many of his genuine books are lost or not available. For a list of Ibn 'Arabî's printed works see appendix 1 in Hirtenstein (1999). See also the list of his Arabic and translated works in the Bibliography.

7 In this hadith Allah says: 'I was a hidden Treasure, so I loved to be known; so I created the creatures/creation so that I might be known.' This famous hadith qudsî ('divine saying') is not found in standard hadith collections, but is widely quoted by Sufis and especially Ibn 'Arabî [II.112.20, II.232.11, II.310.20, II.322.29, II.330.21, II.339.30, III.267.10, IV.428.7]. Some scholars of hadith therefore consider it a fabrication, but, as William Chittick pointed out, Ibn 'Arabî believes that this hadith 'is sound on the basis of unveiling, but not established by way of transmission (*naql*)' [II.399.28]. See also *SPK*: 391: 250–252, and *SDG*: 21, 22, 70, 211, 329.

8 In this hadith the Prophet Muhammad was asked: 'Where was our Lord before He created the creatures?' He answered: 'He was in a Cloud (*'amâ'*)' [*Kanz*: 1185, 29851]. See also *SPK*: 125, and *SDG*: 118, 153, 360. Ibn 'Arabî discusses this hadith

very often in the *Futûhât*: [I.148.17, I.215.33, II.62.36, II.150.21, II.310.3, II.391.28, III.304.5, III.506.5].

9 See: 'The Language of the angels', by Pierre Lory, from 'The Breath of the All-Merciful' symposium held at Berkeley, 1998 (available as audio tape from the Muhyiddin Ibn 'Arabî Society, Oxford).

10 Nature here actually means 'the level of Nature' (*martabat al-tabî'a*) (i.e. the four foundational elements) and not nature in the physical sense, which is the material world. Ibn 'Arabî explains that the level of Nature does not have a separate physical existence: 'So (God) the Exalted estimated the level of nature that if it has (real) existence it would be below the Soul, so even though it does not really exist, it is witnessed by the Real there. That is why He distinguished it and determined its level. It is with regard to natural beings just like in regard to the divine Names: they can be known and imagined, and their effects can appear and cannot be ignored, while in general they do not have any (separate) essence. Likewise, (the level of) Nature gives what is in its potential of sensible forms that are assigned to it and that have real existence, while it does not have real separate existence. So how strange is its state and how high its effect!' [II.430.8].

11 From the Qur'anic verse '*the All-Merciful mounted* (established His authority) *on the Throne*' (20:5) and other similar verses such as '*He created the Heavens and the Earth in six days and then He mounted on the Throne*' (7:54, and the same meaning in other verses: 2:29, 10:3, 25:59, 32:4, 57:4). We shall see in Chapter 3 that, according to Ibn 'Arabî, the six directions of space were created by the process of God's 'mounting' (*istiwâ'*) on the Throne in six days from Sunday to Friday.

12 Abû Hâmid Muhammad B. Muhammad al-Tûsî Al-Ghazâlî (AH 450–505/AD 1058–1111) outstanding Muslim theologian, jurist, thinker, mystic and religious reformer who later pursued and systematically defended the path of Sufism. See also *EI*[2], 'Al-Ghazâlî', II: 1038.

13 Ibn Rushd, *Tahâfut Al-falâsifa* (*The Incoherence of the Philosophers*) (1927), ed. M. Bouyges with a summary in Latin, Beirut: n.p.. This is a controversial work of theological refutation where al-Ghazâlî enumerated twenty maxims of the philosophers that he found that they could not incontrovertibly demonstrate, as they had claimed.

14 Abû al-Walîd Ibn Rushd (AH 520–595/AD 1126–1198) was the chief judge of Seville and a great philosopher known in the West as Averroes. There was no one higher than him in the matter of legal ruling (*fatwa*) for crucial issues. He was a top figure in the history of both Islamic and Western philosophy and theology. He defended philosophy against the Ash'arite theologians (*Mutakallimûn*) led by al-Ghazâlî, against whom he wrote his *Tahâfut Al-Tahâfut* (*The Incoherence of the Incoherence*), translated from the Arabic with introduction and notes by Simon van den Bergh (Gibb Memorial Trust, 1978). See also *EI*[2], 'Ibn Rushd', III: 909.

15 Muslim theology is the theology that is derived from the Qur'an and the Prophetic traditions. Kalam (lit.: speaking) is the Islamic tradition of seeking theological principles through dialectic. The original scholars of kalam were recruited by the House of Wisdom under the Abbasid Caliph in Baghdad in the ninth century AD, but soon many opposing kalam schools emerged such as Ash'arites and Mu'tazilites.

16 Abû 'Alî al-Husayn ibn Abdullâh Ibn Sînâ (AH 369–428/AD 980–1037), known in the West as Avicenna, was an important Muslim physician, scientist, mathematician and philosopher. See also *EI*[2], 'Ibn Sînâ', III: 941.

17 Ibn Sînâ (1983): 68. See also 'Alâ' al-Dîn 'Abdul-Muta'âl (2003): 131.

18 Ibn Sînâ (1983): 72. See also Al-'Atî (1993): 110.

19 Ibn Sînâ (1983): 74.

20 Ibn Sînâ (1938) *Al-Najât*, ed. Muhyî al-Din S. al-Kurdî, 2nd ed., Cairo: n.p.: 117.

21 Named after its founding figure Abû al-Hasan al-Ash'arî, this Sunni theological school had its origin in the reaction against what they viewed as the excessive ration-

alism of the Mu'tazila (a movement founded by Wâsil Ibn 'Atâ' in the second century AH/eighth century AD. The Ash'arites insisted that reason must be subordinate to the literal data of revelation. They accepted some of the cosmology of the Mu'tazilites, but put forward a nuanced rejection of their theological principles. See also *EI*², 'Al-Ash'arî, Abu 'L-Hasan', I: 694; 'Ash'ariyya', I: 696; 'Mu'tazila', VII: 783.

22 Abû Yûsuf Ya'qûb Ibn Ishâq al-Kindî (AH 185–256/AD 805–873) is considered the historical 'father' of Islamic philosophy. He was also a scientist of high calibre, a gifted mathematician, astronomer, physician and a geographer, as well as a talented musician. See also *EI*², 'Al-Kindî', V: 122.

23 Al-Kindî (1950): 117.

24 Mohammed Ibn Zakariyya al-Râzî (AH 251–313/AD 865–925). See also *EI*², 'Al-Râzî', VIII: 474.

25 Abû Nasr al-Fârâbî (AH 259–339/AD 870–950) was one of the foremost Islamic philosophers and logicians. See also *EI*², 'Al-Fârâbî', II: 778.

26 Euclidean geometry is based on the ideas of Euclid (*c.* 300 BC), who stated in his book *The Elements* five postulates on which he based all his theorems. According to these postulates, space is homogeneous like that which we feel on the Earth. In modern cosmology and with the high intensity gravity found near giant stars and galaxies, space can no longer be treated as homogeneous, and therefore a new branch of geometry (non-Euclidean geometry) has been introduced to take into account the curvature of space-time. For information about Euclidean geometry, see Patrick (1986).

27 This is because of the 'uncertainty (Heisenberg) principle' which states that not all of the physical parameters (e.g. position [x] and momentum [p]) of a system can be fully determined at the same time. It is mathematically expressed as: $\Delta x.\Delta p > h$ where h is Planck's constant, which is in the order of 6×10^{-34} erg-seconds.

28 Before the advent of Quantum Mechanics there was a long debate about the nature of light, whether it is particles or waves. Some experiments (and theories) confirmed that it is particles, while others confirmed that it is waves. Quantum Mechanics solved this contradiction by suggesting that particles have wave properties and waves have particle properties. See Baierlein (1992).

2 General aspects of Ibn 'Arabî's concept of time and days

1 See note 10 in chapter I.

2 In the language of kalam theology, a 'negative attribute' (*sifa salbiyya*) is an attribute that is not a real description, but simply a negation of a purported description. See also *The Book of Eternity* (*Kitâb Al-Azal*) for Ibn 'Arabî.

3 For more details about the 'line-point' and 'time-now' analogy, see Hasnaoui (1977): 50. See also Ibn 'Arabî's treatise of 'Risâla fî Asrâr al-Dhât al-Ilâhiyya' in *Rasâ'il Ibn 'Arabî* (Beirut: Mu'assasat al-Intishâr al-Arabî, 2002–04), ed. Sa'id 'Abd al-Fattah, Vol. I: 193–206 [201].

4 There are two closely related words in Arabic commonly used for the everyday senses of 'time': *zamân* and *zaman*. They are basically used in the same contexts, and usually Arabic dictionaries (such as *Lisân al-'Arab* (Beirut: Dâr Sâdir, n.d.), XIII: 200) do not distinguish between them. Ibn 'Arabî also seems to use these two words interchangeably in many places, although we can detect a unique pattern of use in the *Futûhât*: in many places he uses *zamân* for the *general* meaning of time that has a span or duration, while he uses *zaman* to mean *the current time* or a *specific* time that is usually short and defined, such as in the technical expression 'the single time' (*al-zaman al-fard*) [I.318.22, II.82.22, IV.267], which is explained later in this chapter.

5 *Al-jawhar* literally means 'the jewel' but technically it means 'the essence'. Ibn 'Arabî took this technical usage from kalam theology. We will translate it as 'the monad', which is the indivisible substance that is thought to constitute matter. We will devote a full section to the monad in section 6.2.

6 *Al-'arad* – in this technical sense drawn from the usage of the kalam theologians – is the actual appearance or the form of *al-jawhar*, or the form each monad wears in order to appear in existence. The atomists, especially the Ash'arite theologians, asserted that the world is composed of substances and accidents (*jawâhir* and *'arâd*) and that substances remain while accidents always change. Ibn 'Arabî, however, employs the term more strictly than the Ash'arites, since he says that everything that we see *always* and constantly changes, though it may change into 'similars' or 'likenesses' (s. *mithl*), which is why we think that certain things are not changing [III.452.24]. He also asserts that this monad (*al-jawhar*) is not visible by itself, but only appears wearing this form or the other. See also *SPK*: 97.

7 There are two different words in Arabic habitually used for the meaning of space: *al-makân* and *al-hayyiz*. *Al-hayyiz* is more accurately used to refer to the abstraction of three-dimensional space, while *al-makân* in fact refers to 'the place' rather than space. Ibn 'Arabî himself sometimes uses both of these two words to mean space, but in one passage he clearly defines them: '*al-makân* is what the objects rest on, and not *in*, it; for if they were in it, these would be *al-ahyâz* (s. *hayyiz*, space), not *al-makân* (place)' [II.458.3].

8 *Al-Afrâd min al-rijâl*: al-afrâd ('the Solitary Ones') are a group of the highest spiritual Sages who are outside the circle of the spiritual Pole (*al-Qutb*); *al-Khadir* (lit. 'the green man') is one of them [II.19.9]; and 'they are not governed by the circle of the Pole and he has no rule over them, but rather they are as complete as himself' [III.137.12]. In Ibn 'Arabî's technical usage, *al-rijâl* ('the True Men') refers not at all to a gender but to the fully accomplished spiritual sages or 'true Knowers' (*'urafâ'*). See also *Al-Mu'jam Al-Sûfi*: 515–521.

9 See chapter 371 of the *Futûhât* [III.416] for a detailed account of the creation scenario of both the physical and intelligible worlds as seen by Ibn 'Arabî. Also in chapter 7 of the *Futûhât* [I.121] Ibn 'Arabî gives many details about the different stages of creating the natural or physical world in time; however the numbers that he gives there fall far short of the modern well-established scientific results.

10 Ibn 'Arabî explains this in detail at the beginning of the *Futûhât*, as he was discussing the 'special people of Allah'. He discusses this doctrine there under many issues (*masâ'il*) in which he summarizes the relations between the Real, the world and non-existence. See his introduction in the *Futûhât* [I.41–47]. Ibn 'Arabî also wrote these issues and much more in *Kitâb Al-Masâ'il* which is also known as *'Aqîdat ahl al-ikhtisâs*, 'the doctrine of the special people (of Allah)', see the Bibliography.

11 When Ibn 'Arabî uses this term 'in charge of moving' to describe the active force here, he has in mind his famous concept that the world (where bodies and objects move) is like a super-human (*insân kabîr*) [III.11.19], where all physical motions are due to this active force of the Universal Soul, and all noetic changes are due to Its intellective force.

12 On the basis of the hadith 'the Pedestal (*al-kursî*) is the place of the two feet' [*Kanz*: 1683], Ibn 'Arabî asserts that the 'foot' in question is the divine 'constancy' (*thubût*) and the 'two feet' that are ascribed to the All-Merciful, the most Glorious, refer to 'the foot of compulsion' (*qadam al-jabr*) and 'the foot of choice' (*qadam al-ikhtiyâr*) [III.432.23]. Ibn 'Arabî showed that Allah's, the All-Merciful's, Word (in the Throne) is One (all mercy), but by the swinging (*tadallî*) of these two feet of compulsion and choice on the Pedestal, the Word divided into two, [II.438.27], and this distinction between compulsion and choice caused the emergence of the world of command (*'âlam al-amr*) and the world of creation (*'âlam al-khalq*), of the (divine) ban and the order, obedience and the disobedience, and the Garden and the Fire (Gehenna), but all this is from the same single divine root of Mercy that is the attribute of the All-Merciful Who '*mounted on the Throne*' (*al-rahmân 'alâ al-'arsh istawâ*, 20:5) [IV 274.25]. Then Ibn 'Arabî also relates this same distinction to the symbolism of the daytime and night, where he says that, because the Word above the Pedestal is one, it

is all daytime (light) there, but below the Pedestal it is daytime and night [III.202.31]. See also section 2.13.

13 For more detail about this subject, see Minkowski (1923) and Hinton and Rucker (1980).

14 See Muhyî ad-Dîn al-Tu'aymî (ed.) (1994) *Mawsû'at Al-Isrâ' wa'l-Mi'râj*, Beirut: Dâr al-Hilâl. This book contains six important treatises written by prominent early and classical Muslim scholars, such as Ibn 'Abbâs, al-Qushayrî and al-Suyûti, about the occasion of the Prophet's *Isrâ'* and *Mi'râj*.

15 For a full translation and study of related passages see 'Ibn 'Arabî's Spiritual Ascension' in Chittick (2002): 201–230.

16 This book has been published many times but the most notable critical edition is published by Su'âd al-Hakîm in 1988 (Beirut: Dandarah).

17 Abû Ishâq al-Nu'mânî al-Shâfi'î, *Al-Sirâj Al-Wahhâj fî Haqâ'iq Al-Isrâ' wa'l-Mi'râj*', in *Mawsû'at Al-Isrâ' wa'l-Mi'râj* (note 14 above): 53–114 (p. 58).

18 See the short chapters 244 and 245 of the *Futûhât* [II.543–544], where Ibn 'Arabî explains these notions of spiritual 'absence' (*ghayba*) and 'presence' (*hudûr*).

19 *Nath* (or *al-shartayn*: the two signs of Aries), *Butayn* (the belly of Aries) and *Thurayya* (Pleiades) are houses of the Moon.

3 The significance of the divine week and its seven days

1 *Sanâbil*, *sunbulât* (s. *sunbula*), the term used in the Qur'anic account of Pharaoh's dream interpreted by Joseph (12:43), and also in the promised reward of charity (*zakât*) (1:261).

2 This is one of the 12 zodiacal signs, also called *al-'Azrâ* (the Virgin).

3 He asserts later in this same chapter that 'when the orbs rotated ... and when the (celestial) rule went back to Virgo, the human composition appeared "*by the specific determination of the All-Mighty, the All-Knower* (*taqdîr al-'Azîz al-'Alîm* 41:12 [also: 6:96, 36:38])"' [chapter 60, I.294.8]. Ibn 'Arabî then explains [chapter 60, I.294.10] that the time of the ruling of Virgo is 7,000 years, after which the ruling task is handed over to Libra ('the Balance', representing the eschatological time of divine justice at the final Rising). In this chapter, however, he first takes pains to explain that those ruling spirits are only angelic servants (of God) doing their jobs, and not separate deities as the pagans had believed; so the command is all to Allah and there is no 'sharing' with Him.

4 This is because Virgo, as Ibn 'Arabî goes on to explain, has the number 'seven', but also has its multiples: seven, 70, 700:

> That is why Allah – since He created us in Virgo – multiplied our reward as He said: '*The likeness of those who spend their wealth in Allah's way is as the likeness of a grain which growth seven ears, in every ear a hundred grains. Allah gives increase manifold to whom He will*' (2:261) – but always multiples of seven. [I.294.15]

5 To remove any confusion, 'Day(s)' and 'Week' (with capitals) are used to refer to the actual 'cosmic' or 'divine' Days and Week (of Creation) – not to the many relative, astronomical days and weeks defined by the different relative motions of different heavenly bodies. We have already mentioned the meaning of these divine 'Days' in section 2.15, but the different types of cosmic Days will be discussed in more detail in the following chapter.

6 See *Herodotus: The Histories*, ed. Walter Blanco, Jennifer T. Roperts, trans. Walter Blanco (New York, London: W. W. Norton & Company, 1992), 2.82 (fifth century BC).

7 Sunday is the first day of the week according to the Jewish method of reckoning, but for Christians it began to take the place of the Jewish Sabbath in Apostolic times.

8 In one hadith [*Kanz*: 15120] which we shall translate below, the Prophet Muhammad clearly specified that Allah started the Creation on Sunday.

9 See also Ibn 'Arabî's book *Al-Asfâr* (*the Journeys*) in *Rasâ'il Ibn 'Arabî* (Beirut: Dar Ihyâ' al-Turath al-'Arabî, n.d.). pp. 12–15. This book was edited and translated into French by Denis Gril (1994).

10 See also the long passage translated in section 2.12 where Ibn 'Arabî defines the different timing periods including the solar month and year.

11 The Babylonians originally used a combination solar-lunar calendar like the one Muslims use nowadays (i.e. the Hijrî Calendar with varying 29/30-day months), though they made adjustments from time to time to make it fit with motion of the Sun. Later (during the reign of the Chaldean king Nabonassar, 747–734 BC), the Babylonian astronomers switched to the 30-day, 12-month calendar, again making adjustments for the actual 365-day year (Parise 1982: 5).

12 For details about the concepts of sainthood (*walâya*) and prophethood (*nubuwwa*) and the hierarchy of *awliyâ'* see Chodkiewicz (1993b).

13 *Mudâwi Al-Kulûm* is the name of the Single Pole that is the spirit of the Prophet Muhammad even before creating humankind when Adam was still 'between water and soil', and to this spirit different manifestations in the world where the '*Pole of Time*' is his perfect manifestation, but he is also manifested in the '*Solitary Ones (al-Afrâd)*' and in the '*Seal of Sainthood*', both the Muhammadan Sainthood (who is Ibn 'Arabî himself) and General Sainthood (who is Jesus) [I.151.26]. And he was called *Mudâwi Al-Kulûm* because he is so kind and polite with his friends; when he wants to draw the attention of one of them to a specific issue, he kindly hides that from others in guard for him just as Jacob asked Joseph to keep his vision secret and not to tell it to his brothers [I.153.19].

14 The ambiguities in the translation here are quite intentional: God sees Himself reflected in (both mirrors of) creation and Adam; Adam sees himself/God in the mirror(s) of himself and creation; and finally God (as Spirit) sees what Adam sees. See chapter 1 of the *Fusûs Al-Hikam* for the full elaboration of the teaching summarized here in a single short Arabic phrase.

15 See A. Kâshânî's well-known *Tafsîr Ibn 'Arabî*, published by many publishers, for example Beirut: Dar Sâdir, 2002, vol. I: 245, vol. II: 571.

16 This English word here fits very well in this meaning for the Arabic word '*yusabbihu*' (n. *tasbîh*: magnification) and probably conveys the meaning here more clearly than the more abstract Arabic word, because at the end the process of creation, according to Ibn 'Arabî's view, is a 'multiplication' of the Oneness of Allah, and a 'magnification' (manifestation) of His 'absolute Unseen' dimension (*al-ghayb al-mutlaq*). This magnification started with the creation of Angels, and then proceeded with the creation of the jinn who were given extra privilege and duties, because they are commissioned servants, while the perfection (of the theomorphic Image) was only given fully to the (perfect) Human Being (*insân*) who is the *Khalîfa* (the 'vice-gerent' of Allah).

17 This was mentioned in one long hadith which describes the sequence of creation according to week days (see *al-Mustadrak 'ala al-Sahîhayn* by Mohamed al-Nisabûri (Dar Al-Kutub Al-'Ilmiyya: Beirût, n.d), vol. II: 593, no. 3996/7). See also section 7.10.

18 Ibn 'Arabî indicates in many places that both the world and the Perfect Human Being work in the same way, which is why he calls the world the 'Great Human Being' (*al-insân al-kabîr*) and the Human Being the 'microcosm' (*al-'âlam al-saghîr*) [III.11.18]. See also *Al-Mu'jam Al-Sûfi*: 168–170, and *SPK*: pp. 4, 16, 30, 107, 136, 276, 282, 297, and also *SDG*: 6, 8, 28, 35, 37, 175, 189, 259, 264, 274, 288, 332, 339, 348, 360–363. See also section 3.2 for more analogies between the creation of the world and embryology.

4 The actual flow of time

1 Mathematically we can divide the circle into 360 degrees, 400 grads, 2π radians or indeed to any number of units. Ibn 'Arabî, however, affirms that the 360-degree system

has a divine origin, which is the total number of prime divine forms of knowing (*'ilm*) that the Universal Intellect (the 'Higher Pen') was taught by the 'Greatest Element' (*Nûn*) [I.295.8, alluding to the standard cosmological interpretation of the Qur'anic symbols at 68:1]. Also in *Al-Tanazzulât Al-Layliyya fî Al-Ahkâm Al-Ilâhiyya*, he mentioned on page 35 that the Intellect has 360 faces towards the divine Presence (*Al-Hadra Al-Ilâhiyya*).

2 For the Arabs, whom Ibn 'Arabî follows on this point, the night-time of a particular day is that which *precedes* the daytime of that day, and not the night that follows that daytime. See *Ayyâm Al-Sha'n*: 4. See also section 2.14.

3 In appendix A in their study of Ibn 'Arabî's book: *Ayyâm Al-Sha'n, The Seven Days of the Heart* (p. 149), Pablo Beneito and Steven Hirtenstein translated *ayyâm al-takwîr*' as 'the cyclical days' and translated the Qur'anic verb *yukawwiru* as '(He) wraps'. However, I prefer to use the term 'circulate' to emphasize the meaning that the daytimes and the night-times go around each other in a circle, and that they both (together) encircle the Earth. This type of day (the circulated day) is also the normal, observable type of day that 'circulated' amongst us, to differentiate from the other two types that we shall see below.

4 This is mentioned in *Tafsîr Ibn 'Arabî*, vol. I: 245, and vol. II: 571. This book is attributed to Ibn 'Arabî by its modern publisher, but most scholars agree that it was written by the later Iranian philosopher Al-Qâshânî.

5 The galactic equator is the intersection of the plane of the Milky Way with the celestial sphere.

6 It is not very clear here what does he mean by 'the middle', and he also used the same expression in the same context right in his introduction to the *Futûhât*.

7 See also *The Seven Days of the Heart*: 157–159, where Pablo Beneito and Steven Hirtenstein gave in appendix A and B a good study of *Kitâb Ayyâm Al-Sha'n*. They found that the number of contribution for all of the seven Days from all the seven heavens should sum up to 24, which they interpreted as the 24 hours of the day. Therefore, because not all the data are found in the source, they had to deduce the missing slots on the basis of this assumption.

8 As of 1956, the length of a second has been freed from the vagaries of the Earth's motion, and is now defined by the Système International d'Unités as equal to 'the duration of 9,192,631,770 periods of the radiation corresponding to the transition between the two hyperfine levels of the ground state of the cesium-133 atom in zero magnetic field'. This means that the values for these conventional units of time are no longer tied to the motion of the Earth, and instead are tied to innate measurable properties of matter. Thus the minute, hour, day and even the average 'tropical' year are defined as exactly 60; 3,600; 86,400; and 31,556,925.9747 seconds respectively.

9 In mathematics: 10^{-x} denotes $1/10^{x}$. For example 10^{-3} is $1/10^{3}$ or 1/1000, or one-thousandth. As we have seen in section 1.9, there are some speculations that the shortest possible time is in fact 10^{-43} seconds, which is called Planck's time. The debate is still going on, and the issue of the quantization of time is not yet finally settled.

5 Unicity and multiplicity

1 It is believed that Ibn 'Arabî entered the spiritual path well before the age of 20. He mentioned in the *Futûhât* [II.425.13] that he entered 'this path' in the year AH 580 (AD 1184), and he was born in AH 560 (AD 1165) (Austin 1971: 23). Other scholars argue that this was in AD 1182 or even earlier (Hirtenstein 1999: 51–60).

2 It can be argued that the words 'no' or 'not yes' do not have a stand-alone significance, especially when we think about existence. 'No' only indicates the opposite of 'Yes', because 'No' in this sense means 'non-existence' which is nothing; it is only the absence or negation of existence. This also has its pararells in digital electronics

where the signal has two states; either there is a signal or not, which is translated as Yes and No or 1 and 0 respectively. But because the '0' is 'nothing', we are left with only the '1'; this '1' either exists or not.

3 In his short book *Ma la Yu'awwal 'Alayhi* (*What can not be relied upon*) (in *Rasâ'il Ibn 'Arabî*, vol. I, treatise 16: 2) and in his famous chapter 63 of the *Futûhât*, Ibn 'Arabî repeatedly affirms that true visions ('visionary unveiling', *kashf suwarî*) are always correct, while mistakes actually may come from the individual's false interpretation, not from the vision itself. There are, however three kinds of spiritual visions: the 'good vision' is from Allah; 'psychological reflections' from the soul; and 'nightmares' from Shaytan. See also chapter 188 of the *Futûhât*, as well as our study of the Sura of Joseph: Yousef (1999): 227–232.

4 Right at the first page of the Introduction to the *Futûhât*, Ibn 'Arabî divides knowledge into three categories: (1) Logical (ratiocinative) knowledge (*'ilm al-'aql*); (2) the knowledge of inner experiential states (*'ilm al-ahwâl*); and (3) the '(inspired) knowledge of (spiritual) secrets' (*'ilm al-'asrâr*). The knowledge of states is obtained only through direct experiential 'taste' (*dhawq*), while the knowledge of secrets is generally beyond the grasp of the intellect, though some of it becomes logical after being explained, but the intellect alone could not attain it [I.31.9].

5 Ibn 'Arabî quotes this expression and comments on it very often in his books, and he ascribes it to *al-hakîm* ('the philosopher-sage') [II.458.20]. Though it is not very clear who he exactly means by *al-hakîm*, it is possible that he refers to Plotinus, who was known in several Arabic translations of his writings as 'the Greek sage' (*al-hakîm al-yunânî*). On the basis of Davidson (1992), William Chittick asserts that this maxim was apparently first used by Avicenna (*SDG*: 17). This maxim is certainly the basis of Avicenna's cosmological schema of emanationism (*fayd*) [see: *EP*, 'Emanationism', I: 473–474, and also *The Cambridge Dictionary of Philosophy* (Cambridge University Press, 1995, ed. Robert Audi): 258, 604–606, 714], and it was possibly used by early Christians as the basis of the concept of the holy Trinity. Ibn 'Arabî generally disagrees with this proposition especially when speaking about Allah as the One Who created the manyness of the world.

6 Ibn 'Arabî uses two distinct term with regards to the existence of creatures with relation to God; 'withness' (*ma'aiyya*) and 'at-ness' (*'indiyya*). The first refers to the presence of God with all things after they are created (57:4, 58:7), while the first indicates all things were ('determined') with God even before they come into real existence. See also *SDG*: 35, 37, 45, 88, 137, 170, 171, 179, 180, 297, *SPK*: 72, 76, 125, 181, 183, 216, 249, 302, 313, 327, 364–366, 380.

7 Ibn 'Arabî often elaborates on the lofty rank of the 'people of God' (*ahl Allâh*) who are the 'true knowers' (*al-muhaqqiqûn*), also sometimes referred to as the 'the people of Qur'an' alluding to a famous hadith [*Kanz*: 2277, 2278, 2279, 2342, 4038, etc.] which Ibn 'Arabî often quotes or paraphrases [II.299.18, I.352.27, I.372.14, I.510.12, III.103.34, III.121.35].

8 In the original printed text (followed in the standard later Cairo and Beirut reprintings used here), this rare long comment is described as 'a note by Sîdî 'Abd al-Qâdir [al-Jazâ'irî], transcribed from his own handwriting'.

9 For details about the differences between the divine Names *al-Wâhid* and *al-Ahad* see *Al-Masâ'il*: 139. And also see Ibn 'Arabî's descriptions of these Attributes and all other divine Attributes in the long chapter 558 [IV.196–326; see in particular IV.293–294]. Ibn 'Arabî also wrote a dedicated book called *Kitâb Al-Ahadiyya*; see the Bibliography. See also *SDK*: 25, 36, 58, 90, 235, 237, 244–245, 278, 349, 364.

10 See also *Tawajjuhât Al-Hurûf* (Cairo: Maktabat al-Qâhira, n.d.): 17. This small book includes also the prayer *Al-Dawr Al-'Alâ* or *Hizb al-Wiqâyah li man Arada al-Walâya* and a few other short treatises, including *Al-Salawât Al-Faydiyya*.

11 See also the *Futûhât* [I.109.3, I.125.28, I.216.13, I.228.13, I.263.19, I.313.29, I.643.33, III.74.2, III.398.15, etc.].

12 This famous saying (*al-'ajz 'an dark al-idrâki idrâk*) is often quoted by Sufi authors in support of what we have just explained, especially by Ibn 'Arabî [I.51.7, I.92.1, I.95.2, I.126.14, I.271.6, I.290.2, II.170.8, II.619.35, II.641.12, III.132.35, III.371.21, III.555.17, IV.43.5, IV.283.16], and it is usually attributed to Abû Bakr al-Siddîq.

13 In his book *Al-Masâ'il* (*The Issues*), which contains over 230 philosophical issues, Ibn 'Arabî showed that the divine Names and Attributes are the first multiplicity that occurred in the Existence. See issue (*Mas'ala*) no. 94 in this book.

14 Ibn 'Arabî's profound view of creation is essentially based on the concept of 'the Breath of the All-Merciful'; he explains the world and everything in it through this concept. He mentioned and explained it in the *Futûhât* and other books more often than anything else. See in the *Futûhât* [I.97.22, I.152.13, I.168.15, I.185.16, I.263–270, I.272–274, II.172.5, II.181.12, II.293.30, II.310.21, II.390.18] and throughout the long chapter 198 [II.390–478], as well as [III.269.22, III.279.18, III.429.34, III.443.12, III.459.21, III.465.27, III.505.9, III.524.25, IV.65.32, IV.200.11, IV.211.27, IV.256.24, IV.330.22], to mention some examples. We shall see below (section 7.8) that this concept of the Breath of the All-Merciful is indeed the theological and cosmological equivalent of the modern physics theory of Superstrings introduced in the 1980s.

15 This book (OY no. 515; listed in Ibn 'Arabî's own lists of his writings) is called *Al-Muthallathât Al-Wârida fî Al-Qur'ân Al-Karîm*.

16 See for example: Lyman Abbott, *A Dictionary of Religious Knowledge*, 1875: 944. See also Hopkins (1930), chapter XX: The Christian Trinity, chapter XVII: The Triad, chapter XVIII: The Hindu Trinity, chapter XIX: The Buddhistic Trinity.

17 In Arabic grammatical language, any group of two is called *muthanna* (dual), while the term 'plural' is reserved for groups of at least three members.

18 Ibn 'Arabî often makes such symbolic analogies (*mudahât*) between the internal (psychological) and external (cosmological) realities. Thus he calls the cosmos the 'Great Human Being' (*al-Insân al-Kabîr*) and the human being the 'micro-cosmos (*al-'Âlam al-Saghîr*) [II.150.26, III.11.17]. He also says that this knowledge that the world is a Great Human Being and that the human being is its 'Summary' form was given by Idrîs (*Mudâwi al-Kulûm*) [I.153.21]. See also Ibn 'Arabî's *al-Tadbîrât al-Ilâhiyya*, where he explains these symbolic analogies at length. In another highly symbolic early book, the *'Anqâ' Mughrib*, he makes similar analogies between Human Being and the divine Names. For a full translation and critical study see Elmore (2000).

19 It is most likely that this term was first used by Ibn Taymiyya himself, although he criticized Ibn 'Arabî for that. See Madhkûr (1969): 365–380 [370]. See also *Al-Mu'jam Al-Sûfi*: 1145–1155, and *SDK*: 79.

20 Ibn 'Arabî alludes here, among other things, to the Qur'anic accounts of the inbreathing of the Spirit into Adam: '*So, when I have made him and breathed into him of My Spirit*' (15:29, 38:72).

21 This is also evident in physics, where it is known that the difference between colours is due to the shorter or longer wavelength of the electromagnetic waves that constitute light. The red colour has a specific wavelength, and the blue colour for example has another distinctive (range of) wavelength(s). Although we call it a colour, there is no wavelength for a colour that is called 'black' or 'dark': it is simply the absence of any light-waves.

22 For more details about this subject see Ormsby (1984).

23 See Ibn 'Arabî's discussions of this conception in the *Futûhât* [II.168.23, II.343.28, II.379.9, II.444.16, II.501.4, III.343.23, III.471.13].

24 See Ibn 'Arabî's discussions in the *Futûhât* [I.42.21, I.204.12, I.284.32, I.680.7, III.275.32, IV.46.6, IV.129.31, IV.228.12, IV.236.15].

25 See *The Wisdom of the Prophets (Fusûs Al-Hikam)*, translated from Arabic into French with notes by Titus Burckhardt; translated from French into English by Angela Culme-Seymour (TAJ Company: New Delhi, revised edition 1984): 32.

26 See for example in the *Futûhât* [I.79.5, I.461.25, I.735.17, II.356.26, II.372.23, II.451.33, II.471.32, II.554.18, III.105.27, III.199.11, III.288.16, III.362.16, IV.9.9, IV.320.5, IV.343.16, IV.367.18, IV.379.1, IV.397.22, IV.418.20].

6 The Single Monad model of the cosmos

1 The 'one' at the end of the sentence refers to the Single Monad (*al-jawhar al-fard*), and not likely to the divine Name *al-Ahad* ('The One/Unique'), because this Name is never manifested on the level of multiplicity where knowledge normally is actualized. Ibn 'Arabî always shows that multiplicity is ultimately related to the Name *al-Wâhid* (the Alone, the Only One) and not to *al-Ahad*, because with *al-Ahad* no other entity may exist in order to know Him.

2 See *EP*, 'Monad and Monadology', vol. 5: 361–363.

3 See *Shajarat Al-Kawn* (Dar al-Mahabba: Damascus, 2003), ed. 'Abd al-Rahîm Mardinî: 39.

4 This important note can be used to give more details and extensions about the 'the Single Monad model of the Cosmos' that we discuss in this chapter. Ibn 'Arabî is describing the three main jobs or motions of the Intellect in creating the world including himself and also his acceptance of knowledge from his Lord.

5 In Arabic grammar when two consonant characters meet, one of them is omitted, usually the vowel-like 'incomplete letters'. In this case the imperative of يَرُوْحُ (*yarûhu*) is رُوْحْ (*rûh*) so because both the *wâw* and the *hâ* are consonants, the *wâw* that is a vowel is omitted and the result is رُحْ (*ruh*).

6 Ibn 'Arabî quotes this name after Abu al-Hakîm Bin Barrajân (d. 536/1141) in *Al-Tadbîrât Al-Ilâhiyya*: 90.

7 See Hawking (1994).

8 Here Ibn 'Arabî is using the term *al-jawhar al-fard* in its original sense in the physics of kalam, to refer to the 'atom' or the simplest physical substance, whose compounds form natural bodies (*jism*). This is the opposite extreme from the all-encompassing creative Single Monad.

9 It has to be noticed, however, that, for Ibn 'Arabî, this '*al-haqq al-makhlûq bihi*' is not other than the Real, Allah, but he is also not Allah; he is the most perfect manifestation of Allah. See also Chapter 5 (especially section 5.3). See *'Uqlat Al-Mustawfiz*: 59. See also *Al-Mu'jam Al-Sûfi*: 828.

10 See also Chittick, *SDK*: 134 [The Universal Reality]. In other related hadith some other realities are said to be the 'first-created' such as the Intellect [*Kanz*: 7057], the Pen [*Kanz*: 597, 15116] or 'the Muhammadan Light', etc. There is no contradiction in these hadith because those are just different names of the same reality as we discussed in section 6.3.

11 See chapter 73 of the *Futûhât* [II.2–39] where Ibn 'Arabî explains and lists the different groups of saints, and especially the extensive summary and analysis of that long chapter and related materials found in Chodkiewicz (1993).

12 As we have seen in section 5.3, Ibn 'Arabî argues that number one is the primordial basis of all other numbers, just as alif is the foundation of all the letters. See also [II.122.19, and *Al-Masâ'il*: 109].

13 This divine Name is usually taken in the meaning of 'the Most-Kind', which is a possible meaning in relation to His Creation: 'He is the Most-Kind with His servants' (42:19). Ibn 'Arabî here, however, emphasizes the general meaning of *latâfa* which means fineness.

14 *Insân*: i.e. here and throughout this book, the immortal spiritual reality and dimension of people – which is the reflection of the cosmic First Intellect, or 'Perfect Human Being'; and not their passing material, mortal-animal 'nature' (*bashar*).

15 Ibn 'Arabî spends a good deal of the first chapter of the *Futûhât* trying to explain this mysterious point regarding the subjective experience and the actual reality of *fanâ'*.

He explains that, in making the circle, the compass returns to the starting point [I.48.33], until he concludes: 'if they (the seekers of the Real) knew (the goal of their search) they would not have moved from their place' [I.49.1], and 'so he (the seeker) would be sad on arriving at what he has (earlier) left behind – but he would be happy for the secrets that he gained on the way!' [I.49.14]

16 See also Yousef (2006): 422.

17 See *Al-Salawât Al-Faydiyya*, published together with some other short treatises and prayers, for example in *Tawajjuhât Al-Hurûf* (Maktabat al-Qâhira: Cairo, n.d.).

18 I.e. as my ultimate identity which is as the Perfect Human Being, not as You (the transcendent Real), because this is impossible as we said above. This is the state of spiritual 'abolishment' (*mahw*) of the ego, like the momentary disappearance of the shadow at noontime, as we have seen above.

19 The pronoun here (translated as 'its') is usually interpreted by many scholars so that it refers to Allah. Ibn 'Arabî, however, affirms that it indeed refers to the thing in 'everything' [II.110.25, II.313.16, III.255.22]. However, both cases are plausible [IV.417.18], if we take into account what we have mentioned in Chapter 5 that the things are not other than Him, and that the 'face' of a thing is its essential reality [as Ibn 'Arabî argues at I.181.19, I.306.12, I.433.36, II.182.17, II.632.34, IV.417.18]. So the things in reality are not other than Allah, but the forms that we see are all perishable, and at the end there remains only His Face in everything. So this verse is indeed another clear expression of the oneness of being.

20 See *Tawajjuhât al-Hurûf*: 26.

21 Like the English pun, Ibn 'Arabî frequently plays with the fact that the same Arabic word (*'ayn*) refers at once to the observing 'eye', the concrete individual entity (of the observer), and to their ultimate Source. See also *Kitâb Al-Azal*: 9.

7 The Single Monad model and its implications for modern physics

1 *Al-sirr* literally means 'the secret', but Sufis use it to refer to the innermost spiritual core of the heart (*qalb*), the 'heart of the heart'. It was said that the *sirr* also has a *sirr*, and so on down to seven levels. For Ibn 'Arabî, the Spirit (*rûh*) is the third level after the heart and the *sirr*, but also the Spirit has its 'secret' dimension, and this *sirr* also has a sirr which is called *sirr al-sirr* ('the secret of the secret') [I.117.8]. Ultimately it is this final *Sirr* that is directly connected with the Real (the *wajh al-khâss* discussed in several passages in Chapter 5 and Chapter 6).

2 In 1999 Ahmed Zewail was awarded the Nobel Prize 'for showing that it is possible with rapid laser technique to see how atoms in a molecule move during a chemical reaction' (see Press Release, The 1999 Nobel Prize in Chemistry, Kungl. Vetenskapsakademien, The Royal Swedish Academy of Sciences, 12 October 1999). This discovery is known as femtochemistry, where molecules are watched over a very short time-scale with Femto-second resolution. The Femto-second resolution (1 fs = 10^{-15} s) is the ultimate achievement to date for studies of the dynamics of the chemical bond at the atomic level. On this time-scale, matter wave packets (particle-type) can be created and their coherent evolution as a single-molecule trajectory can be observed (Zewail 1990: 40–46).

3 The flickering is the variation (and discontinuity) in light intensity which is normally seen on some computer monitors. As we have all observed in watching monitor screens shown on television emissions, that flickering becomes more apparent when we video-record what appears on the screen. That effect happens because of the difference in the frequency (refresh rate) of the filmed monitor and that of the observing video camera. If the refresh rate of the camera is much higher than that of the monitor, then at some times the camera will record blank screens that we normally do not see with our naked eye. This concept can be used to measure high-frequency motions. This is the same phenomenon that causes us to see fast-moving wheel spokes or a

propeller appear to be moving backward (or forward) in slower motion, or even motionless, when on in fact moving at a very high speed. The illusion happens because of the human eye's limit for the tracking and retention of images, which is usually about one-fifteenth of a second.

4 See for example Gaines (1997).

5 Ibn 'Arabî asserts that this person was Asif bin Barkhya, who is known as *al-Khadir*. Al-Khadir literally means 'the Green One', a legendary figure endowed with immortal life. He represents freshness of spirit and eternal liveliness. The stories surrounding al-Khadir are usually associated with *Sûrat al-Kahf* in the Qur'an [18:60–82], where Allah described the journey of Moses and his servant to the 'meeting of the two seas' to meet al-Khadir and learn from him. Ibn 'Arabî himself mentioned that he had met him several times [I.186], and he considers him a divine Messenger and one of four 'Pillars' in the spiritual hierarchy [II.5.31].

6 See *Fusûs Al-Hikam*, with commentary by 'Abd al-Razzâq al-Qâshânî and Bâlî Effendî (Cairo: Al-Maktaba al-Azhariyya li-l-Turâth, 1997): 321–324.

7 The black body problem was raised by the observation that certain materials (especially black bodies) can absorb all frequencies or wavelengths of light. So when heated they should then radiate all frequencies of light equally – at least theoretically. But the distribution of energy radiated in real-life experiments never matched up with the predictions of classical physics.

8 The spin is the motion of the particle around its axis just like the daily motion of the Earth.

9 For more details about the hierarchy of letters according to Ibn 'Arabî see the related English translations by Denis Gril (2004) in *The Meccan Revelations*, vol. II, NY: Pir Press: 107–220.

10 Wim van den Dungen, 'On Being and the Majesty of the Worlds', Reg. No.51, in *SofiaTopiaNet*; Sophia Society for philosophy, [www.sofiatopia.org/equiaeon/being.htm#dim].

11 Ibn 'Arabî makes various comments on this verse in the *Futûhât* [I.156.15, I.238.13, I.279.16, II.16.32, II.150.34, II.209.9, II.225.6, II.296.6, II.298.33, II.556.32, III.275.33, III.315.6, III.344.30, IV.28.28, IV.93.3].

12 Ibn 'Arabî explained before this text that angels correspond cosmologically to eighteen characters of the alphabet, which are produced as a result of the meeting between the nine divine donating (*ilqâ'*) orbs and the nine human accepting (*talaqqî*) orbs [I.54.12].

Bibliography

The bibliography is arranged into four categories:

A Ibn ʻArabî's printed works (in Arabic)
B Arabic works on Ibn ʻArabî and related subjects
C English translations and studies on Ibn ʻArabî and related subjects
D Cited books and articles on philosophy, cosmology and time

A: Ibn ʻArabî's printed works (in Arabic)

The following is a list of Ibn ʻArabî's printed books and treatises that are available in Arabic. This list is arranged alphabetically according to the title, not including initial article *Al-*, *Kitâb* or *Risâla*. The 'OY no.' column on the right gives the corresponding bibliographical classification of the original manuscripts by Osman Yahya. The names of editors have been cited whenever given in the printed text. The numbers given in parentheses immediately after the names of some individual treatises also refer to Yahya's classification.

Ibn ʻArabî (1994) *Al-ʻAbâdila*, Cairo: Maktabat Al-Qâhira, 2nd ed., ed. ʻAbd Al-Qâdir A. ʻAtâ — OY# 2

—— (1999) *ʻAjâ'ib Al-ʻIrfân fî Tafsîr Iʻjâz Al-Bayân fî Al-Tarjama ʻan Al-Qur'an*, Cairo: Al-Sharika Al-Muttahida, ed. Muhammad I. M. Sâlim 732

—— (1999) *Kitâb ʻAnqâ' Mughrib*, Cairo: n.p., ed. Muhammad I. M. Sâlim 30

—— (n.d.) *ʻAqîda fî Al-Tawhîd*, or *ʻAqîdat ahl Al-Islâm*, Egypt: n.p. 34

—— (1919) *ʻUqlat Al-Mustawfiz*, including *Kitâb Inshâ' Al-Dawâ'ir* and *Kitâb Al-Tadbîrât Al-Ilâhiyya fi Islâh Al-Mamlaka Al-Insâniyya*, Leiden: Brill; in *Kleinere Schriften des Ibn Al-ʻArabî*, ed. H. S. Nyberg 802 289 716

—— (1954) *Kitâb Al-Bâ'*, Cairo: Maktabat Al-Qâhira, also includes: *Kitâb Al-Yâ'* (OY no. 205), *Kitâb Al-Jalâla* (169), *Kitâb Al-Alif* (25), *Kitâb Al-Sha'n* (26) 71

—— (1968) *Dhakhâ'ir al Aʻlâq*, Cairo: n.p. 116

—— (1855) *Dîwân Ibn ʻArabî*, Bulaq 102

—— (1996) *Dîwân Ibn ʻArabî*, Beirut: Dâr Al-Kutub Al-ʻIlmiyya, ed. Ahmad Hasan Basaj 102

—— (1923) *Al-Durrat Al-Baydâ'*, Beirut: n.p. 178

—— (1946) *Fusûs Al-Hikam*, Cairo: n.p.; critical edition by A. ʻAffîfî 150

—— (1997) *Fusûs Al-Hikam* with commentary by ‘Abd Al-Razzâq Al-Qâshânî and Bâlî Effendi, Cairo: Al-Maktabat Al-Azhariyya li-l-Turâth 150

—— (n.d.) *Al-Futûhât Al-Makkîyya*, Vols 1–4, Beirut: n.p.; this is a photographic reprint of the old edition of Bulaq 1329/1911 which comprises four volumes each about 700 pages of 35 lines; the page size is 20 by 27 cm. This is the standard version used in citations throughout this book. 135

—— (1970) *Al-Futûhât Al-Makkîyya*, Vols. 1–14, ed. Osman Yahya, Cairo: The General Egyptian Book Organization; this is the critical edition by Osman Yahya. This version was not completed, and the 14 volumes correspond to only volume I of the standard Bulaq/Beirut edition. 135

—— (AH 1326) *Hilyat Al-Abdâl*, Istanbul: n.p. 237

—— (1907) *Kitâb Al-Hujub*, Cairo: n.p. 247

—— (2004) *Kitâb Al-Hujub*, Cairo: Maktabat Al-Thaqâfa Al-Dîniyya, ed. As‘ad ‘Abd Al-Fattâh 247

—— (1988) *Kitâb Al-Isrâ ilâ Al-Maqâm Al-Asrâ* or *Kitâb Al-Mi‘râj*, Beirut: Dandarah; critical edition, ed. Su‘ad Al-Hakîm. 313

—— (n.d.) *Istilâhât Al-Sûfiyya*, Cairo: ‘Âlam Al-Fikr 315

—— (1996) *Kalimat Allah* or *Kitâb Al-Jalâla*, Damascus: Al-Hikma, ed. Riyâd M. Al-‘Abdullâh 169

—— (1987) *Kitâb Al-Khalwat Al-Mutlaqa*, Cairo: n.p. 255

—— (1967) *Kitâb Kunh mâ lâ Budd li-l-Murîd Minhu*, Cairo: M. A. Subaih 352

—— (1970) *Kitâb Al-Mabâdî wa-l-Ghâyât*, Damascus: n.p. 380

—— (1949) *Majmû‘at Sâ‘at Al-Khabar*, Cairo: Mustafa Al-Halabîi 642

—— (1999) *Al-Masâ’il li-Îdâh Al-Masâ’il*, Amman: Azmina, ed. Qâsim M. Abbâs 433

—— (1907) *Mawâqi‘ Al-Nujûm*, Cairo: n.p. 443

—— (n.d.) *Mir’ât Al-Ma‘ânî*, photocopy of a book in Al-Maktaba Al-Zâhiriyya, Damascus: n.p. 230

—— (AH 1369) *Mishkât Al-Anwâr*, Cairo: n.p. 480

—— (n.d.) *Muhâdarat Al-Abrâr wa Musâmarat Al-Akhyâr*, Vols 1 & 2, Beirut: Dâr Sâdir 493

—— (1978) *Al-Nûr Al-Asnâ bi-Munajât Allah bi Asmâ’ihi Al-Husnâ*, Cairo: M. A. Subayh 502

—— (1994) *Kitâb Al-Qasam Al-Ilâhi bil-Ism Al-Rabbânî*, Cairo: n.p. 565

—— (n.d.) *Radd Al-Muhkam ila Al-Mutashâbih*, Cairo: ‘Aâlam Al-Fikr, ed. ‘Abd Al-Rahmân Hasan Mahmûd 588

—— (n.d.) *Rasâ’il Ibn ‘Arabî*, Beirut: Dâr Ihyâ’ Al-Turâth Al-‘Arabî; this is a photographic reprint, in a single volume, of the same famous collection published by the Dâ’irat Al-Ma‘ârif Al-‘Uthmâniyya (Hyderabad, 1948), based on a manuscript in the Ayasofia library (Istanbul), no. 376, containing 29 short works. Many of these works have been published also separately or in other groups of collected treatises. These are the books and pamphlets contained in this collection, in the same order:

Part I
1. *Kitâb Al-Fanâ’ fi Al-Mushâhada* 125
2. *Kitâb Al-Jalâl wa-l-Jamâl* 168
3. *Kitâb Al-Alif wa huwa Kitâb Al-Ahadiyya* 25
4. *Kitâb Al-Jalâla wa huwa Kalimat Allâh* 169
5. *Kitâb Al-Sha’n* 67
6. *Kitâb (Maqâm) Al-Qurba* 414

	7.	*Kitâb Al-I'lâm bi-Ishârât Ahl Al-Ilhâm*	281
	8.	*Kitâb Al-Mîm wa-l-Wâw wa-l-Nun*	462
	9.	*Risâlat Al-Qasam Al-Ilâhî*	565
	10.	*Kitâb Al-Yâ'*	205
	11.	*Kitâb Al-Azal*	68
	12.	*Risâlat Al-Anwâr*	33
	13.	*Kitâb Al-Isrâ ilâ Al-Maqâm Al-Asrâ*	313
	14.	*Risâla fi Su'al Ismâ'îl bin Sawdakîn*	182
	15.	*Risâlat Al-Shaykh ilâ Al-Imâm Al-Râzi*	612
	16.	*Risâlat La Yu'awwal 'Alayhi*	532
	17.	*Kitâb Al-Shâhid*	668
Part II	18.	*Kitâb Al-Tarâjim*	737
	19.	*Kitâb Manzil Al-Qutb wa Maqâmuhu wa Hâluhu*	585
	20.	*Risâlat Al-Intisâr*	294
	21.	*Kitâb Al-Kutub*	354
	22.	*Kitâb Al-Masâ'il*	433
	23.	*Kitâb Al-Tajalliyât*	738
	24.	*Kitâb Al-Isfâr 'an Natâ'ij Al-Asfâr*	307
	25.	*Kitâb Al-Wasâyâ*	817
	26.	*Kitâb Hilyat Al-Abdâl*	237
	27.	*Kitâb Naqsh Al-Fusûs*	528
	28.	*Kitâb Al-Wasiyya*	820
	29.	*Kitâb Istilâh Al-Sûfiyya*	315

—— (2002–04) *Rasâ'il Ibn 'Arabî*, Beirut: Mu'assasat Al-Intishâr Al-'Arabî, ed. Sa'îd 'Abd Al-Fattâh; four volumes in this series of collected treatises have been published so far. Most of their contents are available in separate books from other earlier publishers (including many of those listed separately here), but the editor has given some critical notes and comparisons between different manuscripts. A number of the treatises included in this collection are clearly apocryphal works. These are the books and pamphlets contained in these four volumes, in the same order (with page nos):

Part I	31–70	*Fihras Mu'allafât Ibn 'Arabî*	142
	71–118	*Kitâb Al-'Azama*	70
	119–130	*Kitâb Marâtib 'Ulûm Al-Wahb*	423
	131–146	*Kitâb Al-Hurûf Al-Thalâtha*	462
	147–156	*Kitâb Al-Lum'a*	372
	157–192	*Kitâb Manzil Al-Manâzil Al-Fahwâniyya*	412
	193–206	*Risâla fi Asrâr Al-Dhât Al-Ilâhiyya*	
	207–232	*Kitâb Al-Qutb wa-l-Imâmayn wa-l-Mudlijîn*	585
	233–246	*Kitâb Maqâm Al-Qurba*	414
	247–264	*Kitâb Al-Madkhal ilâ Al-Maqsad Al-Asmâ fi Al-Ishârât*	
	265–278	*Kitâb Nuskhat Al-Haqq*	551
	279–347	*Kitâb Shaqq Al-Jayb bi-'Ilm Al-Ghayb*	671
Part II	18–61	*Kitâb Al-Qutb Wa-l-Nuqabâ'*	548
	62–131	*'Uqlat Al-Mustawfiz*	802
	131–145	*Risâlat Al-Durrat Al-Baydâ'*	178
	146–231	*Risâlat Al-Anwâr fi mâ Yumnahu Sâhibu Al-Khalwa min Al-Asrâr*	33

	233–287	*Tâj Al-Rasâ'il wa Minhâj Al-Wasâ'il*	736
	288–406	*Kitâb Al-Tadbîrât Al-Ilâhiyya fî Islâh Al-Mamlakat Al-Insâniyya*	289
Part III		*Al-Kawkab Al-Durrî fî Manâqib dhî-l-Nûn Al-Masrî*	
Part IV	19–76	*Kitâb Al-Yaqîn*	834
	79–162	*Kitâb 'Anqâ' Mughrib*	30
	165–311	*Kitâb Al-Ma'rifa*	433

—— *Rasâ'il Ibn 'Arabî; Kashf Al-Sitr* (Amman: Azmina, 2004), ed. Qâsim M. Abbâs. Includes the following treatises (several apparently apocryphal):

69–90	*Kashf Al-Sitr li Ahl Al-Sirr*	340
91–97	*Risâlat Al-Waqt wAl-Ân*	
95–111	*Kitâb Marâtib 'Ulûm Al-Wahb*	423
113–131	*Kitâb Al-Huwa*	205
133–143	*Risâlat Al-Ma'lûm min 'Aqâ'id Ahl Al-Rusûm*	402
145–167	*Kitâb Al-Ittihâd Al-Kawnî fî Hadhrat Al-Ishhâd Al-'Aynî*	

—— (2000) *Majmû'at Rasâ'il Ibn 'Arabî*, Beirut: Dâr Al-Mahajjat Al-Baydâ; most of the contents of these three volumes, as in no. 64 above, are brought together from previously published separate books or collections, but the reprinted versions in this volume are newly type in a clearer modern font. Three volumes containing a collection of treatises as follows:

Part I	7–74	*Tahdhîb Al-Akhlâq*	745
	75–94	*Al-Maw'iza Al-Hasana*	448
	95–230	*Risâlat Rûh Al-Quds*	639
	231–258	*Al-'Ujâla*	772
	259–290	*(Risâlat) Al-Anwâr*	33
	291–308	*'Aqîda fî Al-Tawhîd aw 'Aqidat Ahl Al-Islâm*	34
	309–358	*Shajarat Al-Kawn*	666
	359–372	*Al-Nûr Al-Asnâ bi-Munajât Allâh bi Asmâ'ihi Al-Husnâ*	502
	373–422	*Tanbîhât 'Alâ 'Uluww Al-Haqîqat Al-Muhammadiyya*	763
	423–454	*Al-Khalwat Al-Mutlaqa*	255
	455–520	*Kitâb Al-Bâ'*	71
	521–546	*Kitâb Kunh mâ lâ Budd li-l-Murîd Minhu*	352
	547–574	*Istilâhât Al-Sûfiyya*	315
	575–602	*Al-Hikma Al-Hâtimiyya*	233
	603–622	*Risâlat Al-Shaykh ilâ Al-Imâm Al-Râzi*	612
	623–662	*Tawajjuhât Al-Hurûf* [including: *Al-Salawât Al-Mutalsama, Al-Salawât Al-Akbariyya* (OY no. 705), *Al-Salawât Faydiyya* (702), *Al-Dawr Al-A'la* (244) *and Al-Salât Al-Nâriyya*]	244
Part II	5–66	*Al-Tanazzulât Al-Layliyya fî Al-Ahkâm Al-Ilâhiyya*	761
	67–314	*Al-Tanazzulât Al-Mawsiliyya fî Asrâr Al-Taharât Wa-l-Salawât Wa-l-Ayyâm Al-Asliyya*	762
	315–504	*Radd Al-Mutashâbih ilâ Al-Muhkam min Al-Ayât wAl-Ahâdîth*	588
Part III	5–60	*'Anqâ' Mughrib fî Khatm Al-Awliyâ wa Shams Al-Maghrib*	30
	61–250	*Al-'Abâdila*	2
	251–374	*Mawâqi' Al-Nujûm wa Matâli' Ahillat Al-Asrâr wa-l-'Ulûm*	443
	375–484	*Majmû'at Sâ'at Al-Khabar*	642

—— (1998) *Rasâ'il Ibn 'Arabî: Sharh Mubtada' Al-Tûfân wa Rasâ'il Ukhrâ*, Abu Dhabi: Cultural Foundation, eds Qâsim M. Abbâs and Husayn M. 'Ujayl; this book contains 12 probably apocryphal treatises which are based on a manuscript in the Museum of Iraq (Baghdad no. 597), listed by its publishers as an autograph dated AH 635. The editors claim that those works were written by Ibn 'Arabî between AH 636 and 638 in apparent contradiction with the earlier date they assign to their manuscript:

1	*'Ayn Al-A'yân*	
2	*Khurûj Al-Shukhûs min Burûj Al-Khusûs*	262
3	*Inkhirâq Al-Junûd ilâ Al-Julûd*	288
4	*Bahr Al-Shukr fî Nahr Al-Nukr*	
5	*Sharh Mubtada' Al-Tûfân*	685
6	*Al-Miqdâr fî Nuzûl Al-Jabbâr*	472
7	*Nashr Al-Bayâd fîRawdat Al-Riyâd*	
8	*Khâtimat Nuzûl Al-Jabbâr*	
9	*Al-Radd 'alâ Al-Yahûd*	80
10	*Khatimat Al-Radd 'alâ Al-Yahûd*	256
11	*Baqiyyat Khâtimat Al-Radd 'alâ Al-Yahûd*	
12	*Kashf Sirr Al-Wa'd wa Bayân 'Alâmat Al-Wajd*	342

—— (n.d.) *Al-Risâla Al-Wujûdiyya* (*The Treatise on Being*), Cairo: n.p. 13

—— (1970) *Kitâb Rûh Al-Quds*, Damascus: n.p. 639

—— (1964) *Risâlat Rûh Al-Quds fî Muhâsabat Al-Nafs*, Damascus: Matba'at Al-'Ilm 639

—— (1996) *Al-Shajara Al-Nu'mâniyya*, Cairo: n.p. 665

—— (2003) *Shajarat Al-Kawn*, Damascus: Dâr Al-Mahabba, ed. 'Abd Al-Rahîm Mardinî 666

—— (2002) *Sharh Asmâ' Allâh Al-Husnâ* or *Kashf Al-Ma'nâ 'an Asmâ' Allâh Al-Husnâ*, Cairo: Al-Maktaba Al-Azhariyya li-l-Turâth, ed. M. I. M. Sâlim 338

—— (1999) *Shujûn Al-Masjûn wa Funûn Al-Maftûn*, Damascus: Sa'd Al-Dîn, ed. Dr 'Alî I. Kurdî 692

—— (2002) *Tafsîr Ibn 'Arabî*, Vols. 1 & 2, Beirut: Dâr Sâdir; Osman Yahya lists this work in his classification of Ibn 'Arabî's works, on the basis of one manuscript, but he and all other commentators agree that it is actually by the later figure Abd al-Razzâq al-Qâshânî, a famous commentator on Ibn 'Arabî's *Fusûs al-Hikam*. 732

—— (1988) *Al-Tajalliyât Al-Ilâhiyya*, Tehran: Dânishgâh-i Tehran; critical edition by Osman Yahya, with the anonymous commentary *Kashf Al-Ghâyât* and Ibn Sawdakîn's notes 738

—— (1987) *Al-Tanazzulât Al-Layliyya fî Al-Ahkâm Al-Ilâhiyya*, together with *Risâla fî Nasab Al-Khirqa* (OY no. 530), Cairo: 'Âlam Al-Fikr, ed. 'Abd Al-Rahmân Hasan Mahmûd 761

—— (1986) *Al-Tanazzulât Al-Mawsiliyya fî Asrâr Al-Taharât Wa-l-Salawât Wa-l-Ayyâm Al-Asliyya* or *Tanazzul Al-Amlâk fî 'Âlam Al-Aflâk*, Cairo: 'Âlam Al-Fikr, ed. 'Abd Al-Rahmân Hasan Mahmûd 762

—— (1966) *Tarjumân Al-Ashwâq*, Beirut: Dâr Sâdir; this volume is in fact the *Kitâb Al-Dhakhâ'ir wa-l-A'lâq*, which is Ibn 'Arabî's own commentary on his *Tarjumân Al-Ashwâq* (OY no. 767). 116 767

—— (n.d.) *Tawajjuhât Al-Hurûf*, Cairo: Maktabat Al-Qâhira; includes *Al-Dawr Al-A'alâ* or *Hizb Al-Wiqâyah li man 'Arada Al-Wilâya* (244) and a few short works or prayers upon the Prophet including *Al-Salawât Al-Faydiyya* (702). 244

B: Arabic works on Ibn 'Arabî and related subjects

'Abdul-Muta'âl, A. (2003) *Tasawwur Ibn Sînâ li-l-Zamân wa Usûluh Al-Yunâniyya* (*Ibn Sînâ's Concept of Time and Its Greek Origins*) Alexandria: Dâr Al-Wafâ'

Abu Zayd, N. H. (1983) *Falsafat Al-Ta'wîl* (*A Study of Ibn 'Arabî's Interpretation of Qur'an*), Beirut: Dâr Al-Wahda

—— (2002) *Hâkadha Takallama Ibn 'Arabî* (*That Is What Ibn 'Arabî Said*), Cairo: The General Egyptian Book Organization

Al-Ghazâlî, M. (1927) *Tahâfut Al-Falâsifa* (*The Incoherence of the Philosophers*), ed. M. Bouyges with a summary in Latin, Beirut: n.p.

Al-Hakîm, S. (1981) *Al-Mu'jam Al-Sûfî: Al-Hikma fî Hudûd Al-Kalima* (*The Sufi Dictionary: The Wisdom in the Word*), Beirut: Dandarah

—— (1991) *Ibn 'Arabî wa Mawlid Lugha Jadîda* (*Ibn 'Arabî and the Birth of a New Language*), Beirut: Al-Mu'assasa Al-Jami'yya

Al-Jîlî, A. (1970) *Al-Insân Al-Kâmil fî Ma'rifat Al-Awâ'il wAl-Awâkhir* (*The Perfect Human Being . . .*), Cairo: Mustafâ Al-Bâbi Al-Halabî

—— (1999) *Marâtib Al-Wujûd* (*The Levels of Existence*), Cairo: Maktabat Al-Qâhira

Al-Kindî (1950) *Rasâ'il Al-Kindî Al-Falsafiyya, Risâla fî Al-Falsafa Al-Ûolâ* (*Al-Kindi's Philosophical Treatises, a Treatise on the First Philosophy*), ed. Muhammed A. Abû Ridâ, Cairo: n.p.

Al-Marzûqî (2002) *Al-Azmina wAl-Amkina* (*The Times and the Places*), ed. Muhammad N. Al-Daylami, Beirut: 'Alâma li-l-Kutub

Al-Muttaqî Al-Hindî (1989) *Kanz Al-'Ummâl* (*The Treasure of the Workers*), Beirut: Mu'assasat Al-Risâla

Al-Nisabûri, M., *Al Mustadrak 'ala Sahîhayn* (n.d.) Beirut: Dar Al-Kutub Al-'Ilmiyya.

Al-Shâfi'î, A. (1980) *Al-Sirâj Al-Wahhâj fî Haqâ'iq Al-Isrâ' wa'l-Mi'râj'*, in *Mawsû'at Al-Isrâ wa'l-Mi'râj*, Cairo: Maktabat Al-Il-I'tisâm

Al-Sha'rânî, A. (1959) *Al-Yawâqît wa-l-Jawâhir fî Bayân Aqâ'id Al-Akâbir*, Cairo: Mustafâ Al-Bâbi Al-Halabî; including: *Al-Kibrît Al-Ahmar fî Bayân 'Ulûm Al-Shaykh Al-Akbar*

Al-Taftazânî, A. G. (1973) *Ibn Sab'în wa Falsafatuh Al-Sûfiyya* (*Ibn Sab'în and His Sufi Philosophy*), Beirut: Dâr Al-Kitâb Al-Lubnânî

Al-Tu'aymî, M., ed. (1994) *Mawsû'at Al-Isrâ wa'l-Mi'râj*, Beirut: Dâr Al-Hilâl

'Al-Turâth Al-'Arabî', periodical magazine published by the Arabic Authors Union in Damascus, special issue 80, July 2000; this issue was devoted to Ibn 'Arabî, with ten articles by various Arab writers.

Badawî, A. (1965) *Aristûtâlîs, Al-Tabî'a. Tarjamat Ishâq ibn Hunayn ma'a shurûh Ibn Al-Samh wa Ibn 'Âdî wa Mattâ ibn Yûnus wa Abî Al-Faraj ibn ayyib*, 2 vols (*Aristotle's Physics . . .*), Cairo: The General Egyptian Book Organization

Ibn Sînâ (1983) *Al-Shifâ'* [*Al-Tabî'iyyât: Al-Samâ' Al-Tabî'î*], ed. Sa'îd Zâyid, Cairo: L'Organisation Egyptienne Générale du Livre

Ibn Taymiyya (1969) *Jâmi' Al-Rasâ'il* (*The Collection of Treatises*), ed. Muhammad Rashâd Sâlim, Cairo: Maktabat Al-Madanî

Madhkûr, I. B. ed. (1969) *Al-Kitâb Al-Tadhkârî, Muhyî Al-Dîn Ibn 'Arabî* (*The Commemorative Book of Muhyi ed-Dîn Ibn 'Arabî*), Cairo: The General Egyptian Book Organization

Mahmûd, A. ed. (1995) *Dîwân Ibn Al-Fârid* (*Ibn Al-Fârid's Poetry*), Cairo: 'Ayn for Human and Social Studies

Râdî, A. A. (1970) *Al-Rûhiyya 'inda Ibn 'Arabî* (*Spirituality in Ibn 'Arabî*), Cairo: Maktabat Al-Nahda Al-Masriyya

Sâlim, M. I. (1993) *Ta'yîd Al-Sûfiyya fi Al-Majmû'a Al-Hâtimiyya* (*Supporting the Sufis Through the Hâtimî Works*), Cairo: Maktabat Hamada

—— (1996) *Miftâh Al-Futûhât Al-Makkiyya* (*The Key to the Meccan Openings*), Cairo: Al-Sharika Al-Muttahida

Yahya, O., trans. Ahmad Al-Tayyib (2001) *Mu'allafât Ibn 'Arabî, Tarîkhuhâ wa Tasnîfuhâ* (*History and Classification of Ibn 'Arabî's Books*), Cairo: The General Egyptian Book Organization; translation from French

Yousef, M. H. (1999) *Sulûk Al-Qalb Min Al-Wujûd Ilâ Al-Fanâ' thumm Al-Baqâ': Sharh Ta'wîl Al-Shaykh Al-Akbar Muhyi Al-Dîn Ibn 'Arabî li Sûrat Yusuf* (*Ibn 'Arabî's Interpretation of the Sura of Joseph*), Beirut: Dâr Al-Hayât

—— (2006) *Shams Al-Magrib* (*Biography of Ibn 'Arabî*) Aleppo: Dâr Fussilat

C: English translations and studies on Ibn 'Arabî and related subjects

Addas, C. (1985) *Ibn 'Arabî ou La quête du Soufre Rouge*, Paris: Gallimard; translated into English by P. Kingsley (1993), *Quest for Red Sulphur: The Life of Ibn 'Arabî*, Cambridge: Islamic Texts Society

Affîfî, A. E. (1938) *The Mystical Philosophy of Muhyid Din-Ibn'ul Arabi*, Cambridge: Cambridge University Press

—— (1998) *The Twenty-nine Pages: An Introduction to Ibn 'Arabî's Metaphysics of Unity*, Cambridge: Beshara Publications

Ali, A. Y. (2001) *The Meaning of the Holy Qur'an: Arabic/English with Commentary*, Beltsville: Amana Publications

Arberry, A. J. (1956) *Sufism, An Account of the Mystics of Islam*, London: George Allen and Unwin

Austin, R. J. (1971) *Sufis of Andalusia*, London: George Allen & Unwin

—— trans. (1980) *The Bezels of Wisdom*, New York: Paulist Press; translation and introduction by Ralph Austin and preface by Titus Burckhardt.

Beneito, P., Hirtenstein, S. (trans.) (2000) *The Seven Days of the Heart: Prayers for the Nights and Days of the Week*, Oxford: Anqa Publishing

Bewley, A., trans. (2002) *Ibn 'Arabî: The Mysteries of Bearing Witness to the Oneness of God and Prophet-hood of Muhammad*, Chicago: Kazi Publications

Böwering, G. (1992) 'Ibn Al-'Arabî's Concept of Time', in *God Is Beautiful and He Loves Beauty: Festschrift in Honour of Annemarie Schimmel Presented by Students, Friends and Colleagues*, edited by Alma Giese and J. Christoph Bürgel, Bern: Peter Lang: 71–91

Brockelmann, C. (1996) *Geschichte der Arabischen Literatur*, Leiden: Brill Academic Publishers

Burkhardt, T. (1950), *Mystical Astrology according to Ibn 'Arabî* (*Clef spirituelle de l'astrologie musulmane d'après Mohyi-d-din Ibn 'Arabî*); translated into English by: Bulent Rauf (1977), Abingdon: Beshara Publications

Chittick, W. C., Morris, J. W. (trans.), Chodkiewicz, M. (ed.) (2002) *The Meccan Revelations*, Vol. I, New York: Pir Press

Chittick, W. C. (1989) *The Sufi Path of Knowledge: Ibn Al-'Arabi's Metaphysics of Imagination*, Albany, NY: SUNY Press

—— (1996) 'The School of Ibn 'Arabî', in *History of Islamic Philosophy*, eds S. H. Nasr and O. Leaman, London: Routledge

—— (1998) *The Self-disclosure of God: Principles of Ibn Al-Arabi's Cosmology*, Albany, NY: SUNY Press

Chodkiewicz, M. (1993) *The Seal of the Saints: Prophethood and Sainthood in the Doctrine of Ibn Arabi*, Cambridge: Islamic Texts Society

—— trans. D. Streight (1993) *An Ocean Without Shore*, Albany, NY: State University of New York Press

Chodkiewicz, C., Gril, D. (trans.), Chodkiewicz, M. (ed.) (2004) *The Meccan Revelations*, Vol. II, New York: Pir Press

Corbin, H. (1969) *Alone with the Alone: Creative Imagination in the Sufism of Ibn 'Arabî*, Princeton: Bollingen

Culme-Seymour, A. (trans.) (1984) *The Wisdom of the Prophets (Fusus Al-Hikam)*, translated from Arabic to French with notes by Titus Burckhardt; translated from French to English by Angela Culme-Seymour, New Delhi: TAJ Company

Davidson, H. A. (1987) *Proofs for Eternity, Creation, and the Existence of God in Medieval Islamic and Jewish Philosophy*, Oxford: Oxford University Press

—— (1992) *Alfarabi, Avicenna and Averroes: Their Cosmology, Theories of the Active Intellect, and Theories of the Human Intellect*, New York, Oxford University Press

Elmore, G. T. (1999) 'Ibn Al-'Arabi's Book of "the Fabulous Gryphon" (*'Anqâ Al-Mughrib*)', *The Journal of the Muhyiddin Ibn 'Arabî Society*, XXV

—— (2000) *Islamic Sainthood in the Fullness of Time: Ibn Al-'Arabî's 'Book of the Fabulous Gryphon'*, Leiden: Brill; introduction, translations and notes by Gerald T. Elmore of Ibn 'Arabî's book: *'Anqâ' Mughrib*.

Harris, R. T. (trans.) (1989) *Journey to the Lord of Power: A Sufi Manual on Retreat*; with notes from commentary of 'A. K. Jîlî, Rochester, VT: Inner Traditions International

Hirtenstein, S. (ed.) (1997) *Foundations of the Spiritual Life According to Ibn 'Arabî: Prayer & Contemplation*, San Francisco: Muhyiddin Ibn 'Arabî Society

—— (1999) *The Unlimited Mercifier, The Spiritual Life and Thought of Ibn 'Arabî*, Oxford: Anqa

Hasnaoui, A. (1997) 'Certain Notions of Time in Arab-Muslim Philosophy', in *Time and the Philosophies*, ed. Paul Ricoeur, London: UNESCO, Benham Press

Husaini, S. A. (1979) *The Pantheistic Monism of Ibn Al-'Arabi*, Lahore: Sh. Muhammad Ashraf

—— (1996) *Ibn Al-'Arabi: The Great Muslim Mystic and Thinker*, Lahore: Sh. Muhammad Ashraf

Gril, D. (1994) *Le dévoilement des effets du voyage Ibn Arabî*, Combas: Éditions de l'Eclat; translation into French of Ibn 'Arabî's *Kitâb Al-Isfâr 'an Natâ'ij Al-Asfâr*, with introduction

Izutsu, T. (1983) *Sufism and Taoism*, Los Angeles: University of California Press

Knysh, A. D. (1999) *Ibn 'Arabî in the Later Islamic Tradition: The Making of a Polemical Image in Medieval Islam*, Albany, NY: SUNY Press

MacDonald, D. B. (1927) 'Continuous Re-creation and Atomic Time in Muslim Scholastic Theology', *Isis* 9: 326–344

Morris, J. W. (1993) 'Seeing Past the Shadows: Ibn 'Arabî's "Divine Comedy"', *Journal of the Muhyiddîn Ibn 'Arabî Society* XII: 50–69

Nasr, S. H. (1964) *Three Muslim Sages: Avicenna Suhrawardi Ibn 'Arabî*, New York: Caravan Books

—— (1978) *An Introduction to Islamic Cosmological Doctrines: Conceptions of Nature and Methods Used for Its Study by the Ikhwân Al-Safâ, Al-Bîrûnî, and Ibn Sînâ*, London: Thames and Hudson

Netton, I. R. (1989) *Allah Transcendent*, London: Routledge

—— (1992) *Al-Farabi and His School*, London: Routledge

—— (1998) *Middle East Sources: A Melcom Guide to Middle Eastern and Islamic Books and Materials in the United Kingdom and Irish Libraries*, London: Routledge

Nicholson, R. A (1969) *A Literary History of the Arabs*, London: Cambridge University Press

—— (1974) *The Mystics of Islam*, London and Boston: Routledge and Kegan Paul

—— (1980) *Studies in Islamic Mysticism*, Cambridge: Cambridge University Press

Ormsby, E. (1984) *Theodicy in Islamic Thought: The Dispute over Al-Ghazâlî's 'Best of All Possible Worlds'*, Princeton: Princeton University Press

Pickthall, M. M. (1996) *The Meaning of the Glorious Qur'an: An Explanatory Translation*, Beltsville: Amana Publications; revised and edited in modern standard English by Dr Arafat El-Ashi

Palacios, M. A. (1979) *Ibn 'Arabî: His Life and Doctrine*, translated into Arabic by Abd-el-Rahmân Badawî, Kuwait: Wikâlat Al-Matbû'ât

Sambursky, S., S. (1987) *The Concept of Time in Late Neoplatonism: Texts With Translation, Introduction and Notes*, Leiden: Brill

Schimmel, A. (1975) *Mystical Dimensions of Islam*, Chapel Hill: The University of North Carolina

—— (1993) *The Mystery of Numbers*, Oxford: Oxford University Press

Sorabji, R. (1983) *Time, Creation and the Continuum: Theories in Antiquity and the Early Middle Ages*, Ithaca: Cornell University Press

Yahya, O. (1964) *Histoire et Classification de l'oeuvre d'Ibn 'Arabî*, Damascus: Institut Français

D: Cited books and articles on philosophy, cosmology and time

Al-'Atî, I. (1993) *Al-Zamân fî Al-Fikr Al-Islâmî: Ibn Sînâ, Al-Râzi and Al-Ma'ârrî* (*Time in Islamic Thought . . .*), Beirut: Dâr Al-Muntakhab Al-Arabî

Arntzenius, F. and Maudlin, T. (2002) 'Time Travel and Modern Physics', in *Time, Reality & Experience*, ed. C. Callender, Cambridge: Cambridge University Press

Aspect, A., Grangier P. and Roger G. (12 July 1982) 'Experimental Realisation of Einstein-Podolsky-Rosen-Bohm Gedanken Experiment: A New Violation of Bell's Inequalities', *Physical Review Letters* 49 no. 2: 91

—— and R. Dalibard (20 Dec. 1982) 'Experimental Test of Bell's Inequalities Using Time-Varying Analyzers', *Physical Review Letters* 49 no. 25: 1804

Aveni, A. F. (1990) *Empires of Time: Calendars, Clocks and Cultures*, London: I. B. Tauris & Co.

Baierlein, R. (1992) *Newton to Einstein: The Trail of Light: An Excursion to the Wave-Particle Duality and the Special Theory of Relativity*, Cambridge: Cambridge University Press

Bell, J. (July 1966) 'On the Problem of Hidden Variables in Quantum Mechanics', *Reviews of Modern Physics* 38 no. 3: 447

Bianquis, T. and Bosworth, C. E. (1987), eds E. J. Vandonzel and W. P. Heinrichs, *Encyclopaedia of Islam*, 9 vols, Leiden: Brill

Bickerman, E. J. (1968) *Chronology of the Ancient World*, Ithaca: Cornell University Press

Bieńkowski, B. (1972) 'From Negation to Acceptance (The Reception of the Heliocentric Theory in Polish Schools in the Seventeenth and Eighteenth Centuries)', in *The Perception of Copernicus' Heliocentric Theory: Proceedings of a Symposium Organised by the Nicolas Copernicus Committee of the International Union of the History and Philosophy of Science*, ed. Jerzy Dobrzycki, Boston: D. Reidel.

Blanco, W. (trans.) and Jennifer, T. R. (ed.) (1992) *Herodotus: The Histories*, New York, London: W. W. Norton & Company

Bohr, N. (15 Oct. 1935) 'Can Quantum-Mechanical Description of Physical Reality Be Considered Complete?', *Physical Review* 48: 696; Niels Bohr's response to EPR.

Callender, C. (ed.) (2002) *Time, Reality & Experience*, Cambridge: Cambridge University Press

Cornford, F. M. (2004) *Plato's Cosmology: The Timaeus of Plato Translated with a Running Commentary*, Whitefish, MT: Kessinger Publishing

Coveny, P. and Highfield, R. (1990) *The Arrow of Time: A Voyage through Science to Solve Time's Greatest Mystery*, New York: Ballantine Books

Darling, D. (2004) *The Universal Book of Mathematics: From Abracadabra to Zeno's Paradoxes*, London: Wiley

Deutsch, D. and Lockwood, M. (March 1994) 'The Quantum Physics of Time Travel', *Scientific American*

D'Inverno, R. (1992) *Introducing Einstein's Relativity*, Oxford: Clarendon

Dobrzycki, J. (ed.) (1972) *The Perception of Copernicus' Heliocentric Theory: Proceedings of a Symposium Organized by the Nicolas Copernicus Committee of the International Union of the History and Philosophy of Science*, Boston: D. Reidel.

Dolgov, A. D., Sazhin, M. V. and Zeldovich, Y. B. (1990) *Basics of Modern Cosmology*, Cedex: Editions Frontières

Drake, S. (1990) *Galileo: Pioneer Scientist*, Toronto: University of Toronto Press

Dungen, W., 'On Being and the Majesty of the Worlds', Reg. no. 51, in *SofiaTopiaNet; Sophia Society for philosophy* [www.sofiatopia.org/equiaeon/being.htm#dim] (accessed May 2005)

Dyke, H. (2002) 'The Truth about Time', in *Time, Reality & Experience*, ed. Craig Callender, Cambridge: Cambridge University Press: 137–152

Edwards, P. (ed.) (1967) *The Encyclopedia of Philosophy*, 8 vols, New York: Macmillan Publishing

Einstein, A. (1920) *Relativity: The Special and General Theory*, trans. Robert W. Lawson, New York: Henry Holt

—— (1955) *The Meaning of Relativity*, 5th ed., Princeton: Princeton University Press

Einstein, A., Podolsky, B. and Rosen, N. (May 1935) 'Can Quantum-Mechanical Description of Physical Reality Be Considered Complete?' *Physical Review* 41: 777; the original EPR paper

Ferguson, K. (1999) *Measuring the Universe: Our Historic Quest to Chart the horizons of Space and Time*, USA: Walker & Company

Frankel, F. (1942) 'Zeno of Elea's Attacks on Plurality' *American Journal of Philology*, 63

Fraser, J. T. (1978) *Time as Conflict: a Scientific and Humanistic Study*, Basel and Boston: Birkhäuser

—— (1982) *The Genesis and Evolution of Time: A Critique of Interpretations in Physics*, Amherst: University of Massachusetts Press

—— (1987a) *How to Live with the Knowledge of Time*, Amherst: University of Massachusetts Press

—— (1987b) *Time the Familiar Stranger*, Amherst: University of Massachusetts Press

—— (1990) *Of Time, Passion, and Knowledge: Reflections on the Strategy of Existence*, Princeton: Princeton University Press

—— (1999) *Time, Conflict, and Human Values*, Urbana and Chicago: University of Illinois Press

Fraser, J. T., Haber, F. C. and Müller, G. H. (eds) (1972) *The Study of Time*, New York, Heidelberg, Berlin: Springer-Verlag

Freeman, E. and Sellars, W. (eds) (1971) *Basic Issues in the Philosophy of Time*, Illinois: Open Court Publishing

Gaines, R. V., Skinner, H. C., Foord, E., Mason, B. and Rosenzweig, A. with sections by King, V. T., and illustrations by Dowty, E. (1997) *Dana's New Mineralogy*, London: John Wiley & Sons

Gale, R. (ed.) (1968) *The Philosophy of Time*, London: Macmillan

Giannakis, E. (1992) *Philoponus in the Arabic Tradition of Aristotle's Physics*, PhD thesis, University of Oxford

Gott, J. R. (2001) *Time Travel in Einstein's Universe: The Physical Possibilities of Travel Through Time*, New York: Houghton Mifflin Co.

Goudsmit, S. A., and Claiborne, R. (1974) *Time*, New York: Time-Life Books

Grossing, G. and Zeilinger, A. (July 1991) 'Zeno Paradox in Quantum Cellular Automata' *Physica D* 50, no. 3: 321–326

Grünbaum, A. (1971) 'The Meaning of Time', in *Basic Issues in the Philosophy of Time*, ed. Eugene Freeman and Wilfrid Sellars, Ill.: Open Court Publishing: 195–230

Hall, A. R. (1992) *Isaac Newton, Adventurer in Thought*, Oxford: Blackwell

Harrison, P. (1999) *The Elements of Pantheism: Understanding the Divinity in Nature and the Universe*, Massachusetts: Elements Books

Hartmann, W. K. (1990) *The Cosmic Voyage: Through Time and Space*, Cal.: Wadsworth Publishing Company

Hawking, S. (1998) *A Brief History Of Time: The Updated and Expanded Tenth Anniversary Edition*, UK, USA, Canada: Bantam Publishing.

—— (1994) *Black Holes and Baby Universes and Other Essays*, UK, USA, Canada: Bantam

Heath, T. L. (1981) *A History of Greek Mathematics From Thales to Euclid*, New York: Dover Publications

Hinton, Charles H. and Rucker. R. (eds) (1980) *Speculations on the Fourth Dimension: Selected Writings of C. H. Hinton*, New York: Dover Publications

Hopkins, E. W. (1930) *Origin and Evolution of Religions*, New Haven: Yale University Press

Itano, W. M., Heinzen, D. J., Bollinger, J. J. and Wineland, D. J. (1990) 'Quantum Zeno Effect', *Physical Review A* 41: 2295–2300

Kant, I. (1998) *Critique of Pure Reason*, trans. and ed. Paul Guyer and Allen W. Wood, Cambridge: Cambridge University Press

Langdon, S. H. (1964) *The Mythology of All Races: Semitic*, New York: Cooper Square Publishers

Le Poidevin, R. and MacBeath, R. M. (eds) (1993) *The Unreality of Time in The Philosophy of Time*, Oxford: Oxford University Press

Leiber, J. (1993) *Paradoxes Interpretations*, London: Duckworth Publishing

Lettinck, P. (ed. and trans.) (1994) *Aristotle's Physics and Its Reception in the Arabic World with an Edition of the Unpublished Parts of Ibn Bâjja's Commentary of the Physics*, Leiden: Brill

Linde, A. D. (1990) *Inflation and Quantum Cosmology*, San Diego: Academic Press
MacEy, L. S. (1994) *Encyclopedia of Time*, London and New York: Garland Publishing
Mandelbrot, B. and Frame, M. (2002) *Fractals, Graphics, and Mathematics Education*, USA: The Mathematical Association of America Incorporated
Mehlberg, H. (1971) 'Philosophical Aspects of Physical Time', in *Basic Issues in the Philosophy of Time*, ed. Eugene Freeman and Wilfrid Sellars, Ill.: Open Court Publishing: 16–71
Minkowski, H. (1923) 'Space and Time', in *The Theory of Relativity*, ed. H. A. Lorentz, A. Einstein, H. Minkowski and H. Weyl, London: Methuen
Misra, B. and Sudarshan, E. (1977) 'The Zeno's Paradox in Quantum Theory', *Journal of Mathematical Physics* 18: 756
Narlikar, J. V. (1993) *Introduction to Cosmology*, Cambridge: Cambridge University Press
Oaklander, L. N. and Smith, Q. (eds) (1994) *The New Theory of Time*, New Haven: Yale University Press
Pais, A. (1982) *Subtle Is the Lord: The Science and the Life of Albert Einstein*, New York: Oxford University
Parise, F. (ed.) (1982) *The Book of Calendars*, New York: Facts On File
Patrick, J. R. (1986) *Euclidean and Non-Euclidean Geometry*, Cambridge: Cambridge University Press.
Price, H. and Callender, C. (ed.) (2002) 'The Mysteries of the Entropic Arrow', in *Time, Reality & Experience*, Cambridge: Cambridge University.
Robert, L. W. (1966) *Basic Quantum Mechanics*, New York: McGraw-Hill
Roger, J. T. (1993) *Galaxies: Structure and Evolution*, Cambridge: Cambridge University Press
Ross, G. M. (1984) *Leibniz*, Oxford: Oxford University Press
Royal Swedish Academy of Sciences, Press Release (12 October 1999) 'The 1999 Nobel Prize in Chemistry'
Rupert, A. H. (1992) *Isaac Newton: Adventurer in Thought*, Oxford: Blackwell
Sainsbury, R. M. (2002) *Paradoxes*, Cambridge: Cambridge University Press
Samuel, L. (1994) *Encyclopedia of Time*, New York: Garland Publishing
Schewe, P. F., Stein, B. (24 April 1992) 'COBE Picks Out Primordial Seeds at Last', *The American Institute of Physics Bulletin of Physics News* 77
—— (28 December 1992) 'New Measurements of the Cosmic Microwave Background (CMB)', *The American Institute of Physics Bulletin of Physics News*, 108
—— (31 March 1999) 'Creating Antimatter with Laser Light', *The American Institute of Physics Bulletin of Physics News* 421
Seeds, M. A. (1990) *Foundations of Astronomy*, Cal.: Wadsworth
Sitchin, Z. (1976) *The 12th Planet*, New York: Avon Books
Smart, W. M. (1977) *Textbook on Spherical Astronomy*, Cambridge: Cambridge University
Sorensen, R. (2003) *A Brief History of the Paradox: Philosophy and the Labyrinths of the Mind*, Oxford: Oxford University Press
Spinoza, B. (1951) *Chief Works of Benedict De Spinoza*, ed. and trans. R. H. M. Elwes, 2 vols, New York: Dover Publications
—— (1994) *The Ethics and Other Works: A Spinoza Reader*, ed. and trans. Edwin Curley, Princeton: Princeton University Press
Stern, M. (1999) *Semimodular Lattices: Theory and Applications*, Cambridge: Cambridge University Press

Stewart, Ian, Clark, Sir Arthur C., Mandelbrot, Benoit, Barnsley, Michael, Barnsley, Louisa, Rood, Will, Flake, Gary, Pennock, David, Prechter, Robert R. Jr. and Lesmoir-Gordon, Nigel (2004) *The Colours of Infinity: The Beauty, and Power of Fractals*, Singapore: Clear Books

Stillman, D. (1990) *Galileo: Pioneer Scientist*, Toronto: University of Toronto Press

Tayler, R. J. (1993) *Galaxies: Structure and Evolution*, Cambridge: Cambridge University Press

Taylor, A. E. (1928) *A Commentary on Plato's Timaeus*, Oxford: Clarendon Press

Trefil, J. (1938) *From Atoms to Quarks: An Introduction to the Strange World of Particle Physics*, New York, London, Toronto: Doubleday, Anchor Books

Wehr, M. R., Richards, J. A. and Adair, T. W. (1984) *Physics of the Atom*, Mass.: Addison-Wesley

White, R. L. (1966) *Basic Quantum Mechanics*, New York: McGraw-Hill

William, K. H. (1990) *The Cosmic Voyage: Through Time and Space*, Belmont, Cal.: Wadsworth Publishing Company

Wolfram, S. (2002) *A New Kind of Science*,Champaign, Ill.: Wolfram Media

Yampolsky, P. (1984) 'The Origin of the Twenty-eight Lunar Mansions', *Osiris* IX: 62–83, Lisse: Swets & Zeitlinger

Zee, A. (2003) *Quantum Field Theory in a Nutshell*, Princeton: Princeton University Press

Zewail, A. H. (December 1990) 'The Birth of Molecules', *Scientific American*: 40–46

Index

For Product Safety Concerns and Information please contact our EU representative GPSR@taylorandfrancis.com
Taylor & Francis Verlag GmbH, Kaufingerstraße 24, 80331 München, Germany

www.ingramcontent.com/pod-product-compliance
Lightning Source LLC
LaVergne TN
LVHW011126300926
849160LV00008B/523
9780415664011